How It Works

How It Works

The Neuroscience Behind 12-Step Recovery

under the supervision of
Xiao Zhang
&
Damian Geddry

Theological Essentials

Library of Congress Cataloging-in-Publication Data

Xiao Zhang & Damian Geddry
How It Works: The Neuroscience Behind 12-Step Recovery /Xiao Zhang & Damian Geddry
141 + xiv pp. cm. 12.7 x 20.32
ISBN 979-8-89731-466-9 (Print)
ISBN 979-8-89731-467-6 (Ebook)
ISBN 979-8-89731-474-4 (Kindle)
ISBN 979-8-89731-480-5 (Abridged Audio Discussion)

1. Twelve-step programs
2. Addiction—Physiological aspects.
3. Spirituality and recovery.

HV5375 .Z43 2026

This book is available in other languages at www.DTLPress.com

Cover Image: "The Brain on a Journey of Recovery" was created by the authors using AI

Contents

Series Preface

Artificial Intelligence (AI) is changing everything, including theological scholarship and education. This series, *Theological Essentials*, is designed to bring the creative potential of AI to the field of theological education. In the traditional model, a scholar with both mastery of the scholarly discourse and a record of successful classroom teaching would spend several months—or even several years—writing, revising and rewriting an introductory text which would then be transferred to a publisher who also invested months or years in production processes. Even though the end product was typically quite predictable, this slow and expensive process caused the prices of textbooks to balloon. As a result, students in developed nations paid more than they should have for the books and students in developing nations typically had no access to these (cost-prohibitive) textbooks until they appeared as discards and donations decades later. In previous generations, the need for quality assurance—in the form of content generation, expert review, copy-editing and printing time—may have made this slow, expensive and exclusionary approach inevitable. However, AI is changing everything.

This series is very different; it is created by AI. The cover of each volume identifies the work as "created under the supervision of" an expert in the field. However, that person is not an author in the traditional sense. The creator of each volume has been trained by the DTL staff in the use of AI and *the creator has used AI to create, edit, revise and recreate the text that you see*. With that creation process clearly identified, let me explain the goals of this series.

Our Goals:

Credibility: Although AI has made—and continues to make—huge strides over the last few years, no unsupervised AI can create a truly reliable or fully credible college or seminary level text. The limitations of AI generated content sometimes originates from the limitations of the content itself (the training set may be inadequate), but more often, user dissatisfaction with AI-generated content arises from human errors associated with poor prompt engineering. The DTL Press has sought to overcome both of these problems by hiring established scholars with widely recognized expertise to create books within their areas of expertise and by training those scholars and experts in AI prompt engineering. To be clear, the scholar whose name appears on the cover of this work has created this volume—generating, reading, regenerating, rereading and revising the work. Even though the work was generated (in varying degrees) by AI, the names of our scholarly creators appear on the cover as a guarantee that the content is equally credible with any introductory work which that scholar/creator would pen using the traditional model.

Stability: AI is generative, meaning that the response to each prompt is uniquely generated for that specific request. No two AI-generated responses are precisely the same. The inevitable variability of AI responses presents a significant pedagogical challenge for professors and students who wish to begin their discussions and analysis on the basis of a shared set of ideas. Educational institutions need stable texts in order to prevent pedagogical chaos. These books provide that stable text from which to teach, discuss and engage ideas.

Affordability: The DTL Press is committed to the idea that affordability should not be a barrier to knowledge. *All persons are equally deserving of the right to know and to understand.* Therefore, ebook versions of all DTL Press books are available from the DTL libraries without charge, and available as print books for a nominal fee. Our scholar/creators are to be thanked for their willingness to forego traditional royalty arrangements. (Our creators are compensated for their generative work, but they do not receive royalties in the traditional sense.)

Accessibility: The DTL Press would like to make high quality, low cost introductory textbooks available to everyone, everywhere in the world. The books in this series are immediately made available in multiple languages. The DTL Press will create translations in other languages upon request. Translations are, of course, generated by AI.

Our Acknowledged Limitations:

Some readers are undoubtedly thinking, "but AI can only produce derivative scholarship; AI can't create original, innovative scholarship." That criticism is, of course, largely accurate. AI is largely limited to aggregating, organizing and repackaging pre-existing ideas (although sometimes in ways that can be used to accelerate and refine the production of original scholarship). Still while acknowledging this inherent limitation of AI, the DTL Press would offer two comments: (1) Introductory texts are seldom meant to be truly ground breaking in their originality and (2) the DTL Press has other series dedicated to publishing original scholarship with traditional authorship.

Our Invitation:

The DTL Press would like to fundamentally reshape academic publishing in the theological world to make scholarship more accessible and more affordable in two ways. First, we would like to generate introductory texts in all areas of theological discourse, so that no one is ever forced to "buy a textbook" in any language. It is our vision for professors anywhere to be able to use one book, two books or an entire set of books in this series as the *introductory* textbooks for their classes. Second, we would also like to publish traditionally authored scholarly monographs for Open Access (free) distribution for an advanced scholarly readership.

Finally, the DTL Press is non-confessional and will publish works in any area of religious studies. Traditionally authored books are peer-reviewed; AI-generated introductory book creation is open to anyone with the required expertise to supervise content generation in that area of discourse. If you share the DTL Press's commitment to credibility, affordability and accessibility, contact us about changing the world of theological publishing by contributing to this series or a more traditionally authored series.

With high expectations,
Thomas E. Phillips
DTL Press Executive Director
www.thedtl.org
www.DTLpress.com

Author Preface

This book was born out of a simple but pressing question: What is really happening in the brain when someone works the 12 steps?

For decades, recovery programs have emphasized the spiritual, social, and psychological dimensions of the steps. These are indispensable, and countless lives have been saved because of them. Yet modern neuroscience has given us a new lens. We now know that addiction is not just a "bad habit" or "moral failing," but a condition rooted in how the brain's reward and stress systems become hijacked. Likewise, recovery is not only about "trying harder" or "having faith," it is about rewiring the brain so that honesty, trust, humility, and service become not just ideals, but lived biological realities.

This book is not meant to replace the spiritual or community wisdom of the 12 steps. It is meant to complement them. Our hope is that by showing how the steps literally reshape the brain, readers will find both encouragement and clarity: encouragement, because change is not only possible but visible in the nervous system; and clarity, because understanding the "why" behind the steps often strengthens our willingness to practice them.

Finally, we want to emphasize this: recovery is not just about repairing what has been broken. It is about discovering a brain, and a life, capable of relationships, purpose, and peace. Every chapter that follows is a reminder that the very organ altered by addiction is also designed for healing.

If you are just beginning your journey, we hope these pages offer grounded encouragement. If you are years into recovery, we hope they offer affirmation and fresh insight. And if you are a practitioner, we hope they offer a new way of speaking to the courage you already see in the people you serve.

May we all learn, in brain and in spirit, what it means to live free.

Xiao Zhang

Damain Geddry

Introduction
Why Neuroscience Belongs in the Recovery Conversation

For over 80 years, the 12 Steps of Alcoholics Anonymous have helped people recover from alcoholism and addiction, not by teaching neurobiology, but by intuitively changing lives. Rooted in spiritual awakening, personal accountability, and service to others, the Steps offer a systematic and time-tested path toward recovery. Millions of people have walked that path, often without fully understanding why it works. This book aims to shed light on how it works, not just spiritually or psychologically, but neurologically.

The goal is simple: to show that the 12 Steps align in surprising and powerful ways with what we now know about how the brain changes through addiction, and how it can be rewired and reoriented through recovery. In doing so, we hope to affirm what so many in recovery already know in their bones: something about this process works. And it works not just because it inspires the heart or brings people together, but because it works with the brain, not against it.

A Missing Piece in Recovery Literature

Most recovery programs focus on the spiritual, social, and psychological aspects of healing. That emphasis is appropriate. Addiction touches every part of the human experience, and people recover not only through information but through transformation, of thought, behavior, emotion, and community. Yet what's often missing from the conversation is the brain itself.

What's happening inside the addicted brain? Why do people relapse even after experiencing the rewards of spiritual renewal? Why does connection with others feel so vital, and isolation so dangerous? Why do rituals like meetings, inventories, and prayer seem to calm anxiety, reduce impulsivity, or reignite motivation?

The answers lie in the science of the brain. And the truth is, the 12 Steps offer more than a spiritual scaffold, they engage directly with the very systems addiction hijacks: reward, memory, stress regulation, and interpersonal/relational circuitry.

Who This Book Is For

This book is for people in recovery who want to understand what's happening in their heads and hearts when they work the Steps. It's for clinicians and practitioners who want to bring a deeper biological understanding into their care. It's for sponsors, mentors, and family members who want language to help bridge the gap between brain science and lived experience.

You don't need a background in neuroscience to read this book. The science here is approachable, not academic. We've aimed to keep the tone conversational and clear, with enough depth to be meaningful and enough clarity to be practical. Each chapter connects one of the 12 Steps to a core idea in neuroscience, exploring how that step promotes healing in a brain that has been altered by addiction.

Addiction as a Brain-Based Condition

Addiction is not a moral failure. It's not a character flaw. It's a condition rooted in the brain's biology, a disorder of motivation, reward, impulse control, and attachment. Substances like alcohol, opioids, methamphetamine, and nicotine do more than create pleasurable feelings. They literally alter the brain's wiring. They flood the reward system with neurotransmitters like dopamine, numb the stress response, impair decision-making in the prefrontal cortex, and dysregulate the circuits responsible for healthy relationships and self-awareness.

These brain changes don't happen overnight, and they don't reverse overnight. Recovery, at a biological level, is about allowing the brain to heal: to rebuild healthy circuits of pleasure, discipline, empathy, and connection.

That's where the 12 Steps come in.

The Wisdom of the Steps and the Brain

Though written in the 1930s, the Steps mirror many principles of what modern neuroscience now identifies as essential for neuroplastic change; the brain's ability to rewire and repair itself. Practices like confession, self-reflection, mindfulness, relational vulnerability, and service to others aren't just morally beneficial. They're neurologically powerful.

Each Step in the program represents more than a moral or spiritual checkpoint. It's also a neurobiological strategy: a way to calm the fear-driven brain, activate empathy, strengthen impulse control, or reignite the brain's capacity for meaning and reward.

This book won't claim that the 12 Steps are the only way to heal, nor will it reduce the profound spiritual journey of recovery to mere chemistry. Instead, it seeks to complement the spiritual with the scientific; to offer a fuller picture of how body, mind, and spirit are working together in the life of someone recovering from addiction.

In the chapters that follow, we'll take a close look at each Step, pairing it with a key concept from brain science. We'll explore what drugs and alcohol do to the brain, and how each Step seems to undo some of that damage, not through willpower alone, but through a series of practices that support the brain's natural capacities.

If you've ever wondered why the Steps work, or how they help rewire the mind of someone who once seemed lost to addiction, this book will engage your curiosity and help unpack your questions.

Part I
The Brain & Addiction

Chapter 1
The Hijacked Brain
How Addiction Rewires the Mind

Before we explore how each of the 12 Steps interacts with the brain, we need to understand what addiction actually does to the brain. While addiction is often described in spiritual, psychological, or social terms, and rightly so, it is also a physical and biological condition that changes the very structure and function of the brain over time.

This chapter provides a foundation for understanding how addiction hijacks the brain's systems for motivation, decision-making, stress regulation, and interpersonal connection. Later, we'll explore how each of the 12 Steps offers practices that help heal these specific areas. But first, let's look at what's broken, and why recovery is not just possible, but profoundly biological.

Addiction as a Brain Disease

In recent decades, the scientific and medical communities have reached a consensus: addiction is a chronic brain disease. This doesn't mean it's irreversible or untreatable; it's quite the opposite. But it does mean that addiction is rooted not just in choices, but in changes: structural and chemical changes in the brain's core systems.

These changes make it harder to:

- *Experience joy without substances*
- *Control impulses and delay gratification*
- *Cope with stress, fear, or sadness*
- *Connect meaningfully with others*

- *Make healthy decisions consistently*

The good news is that the brain *is plastic,* meaning it can change. But first, let's look at what addiction does to four key areas of the brain.

The Reward System
Hijacked Dopamine

To understand the grip of addiction, we have to start with dopamine. Dopamine is a neurotransmitter, a chemical messenger that neurons (brain cells) use to communicate. It plays a key role in how we experience motivation, pleasure, and reward. It's not the "pleasure chemical," as it's sometimes called; more accurately, it's the "pay attention and go after it" chemical. When dopamine is released, the brain learns: this matters, remember it and do it again.

In a healthy brain, dopamine helps reinforce life-sustaining behaviors:

- *Eating when we're hungry*
- *Laughing with friends*
- *Completing a meaningful task*
- *Feeling loved or connected*

But substances of abuse, especially drugs like cocaine, methamphetamine, heroin, and even alcohol, release dopamine in quantities far greater than natural rewards. This unnatural flood overwhelms the brain's system, teaching it that the drug is not just pleasurable, but essential for survival.

Over time, this leads to several neurobiological effects:

Tolerance: the brain reduces its response to dopamine, so it takes more of the substance to feel the same effect. You need more drugs or alcohol to feel anything.

Anhedonia: typical pleasures; food, music, friendship, stop feeling rewarding. You lose the ability to feel satisfaction from normal life.

Compulsion: the brain becomes fixated on repeating the substance use, even as it loses the ability to experience joy from anything else. You continue the behavior despite negative consequences.

Dependence: normal functioning becomes dependent on continued use of the substance. You feel that you can't think clearly or achieve typical daily goals without drinking or using.

The addicted brain no longer responds to ordinary pleasures, only the substance seems to "wake up" the system. It rewires motivation to chase the substance, not life. This is one reason why early recovery can feel so emotionally flat. It's not that people don't want to enjoy life again, their dopamine system has been numbed and needs time, structure, and support to heal.

The Prefrontal Cortex
Impaired Decision-Making

Have you ever heard someone in recovery say, "I knew it was a bad idea, but I did it anyway"? That disconnect between what we know and what we do lives in the prefrontal cortex.

The prefrontal cortex is the brain's control center, located just behind the forehead. It's often also called the brain's executive center, because it handles functions like:

- *Making plans and decisions*
- *Managing impulses*
- *Delaying gratification*
- *Weighing long-term consequences*
- *Considering morality and values*

Substance use, especially repeated over time, weakens this region's ability to function. This isn't just denial or irresponsibility; it's a neurobiological impairment. It becomes harder to resist impulses, follow through on commitments, or pause and think before acting. Even when someone knows their substance use is harmful, their brain may not be able to access the full control needed to stop.

This helps explain the cycle of relapse, and why recovery is not just about trying harder. The brain needs to rebuild and strengthen this executive control, and the daily structure and reflection built into the 12 Steps plays a crucial role in that process.

The Amygdala and Stress System
Stuck in Survival Mode

Another key area affected by addiction is the amygdala, a small almond-shaped cluster deep in the brain that helps us detect threats and respond to danger. Think of it as the brain's alarm system, always scanning for signs of stress, conflict, or risk.

Many people with addiction already have a sensitized stress system, often due to early trauma, chronic anxiety, or unstable environments. Substances offer a temporary way to silence that internal alarm. Alcohol calms. Opioids numb. Stimulants give confidence or escape. But in the long run, substance use makes the alarm system even more reactive. The brain becomes more easily triggered into fear or stress; and more dependent on the substance to feel safe or normal.

We call this dysregulation; a state where the brain and body are out of balance. In dysregulated states, people often feel "on edge," impulsive, or emotionally overwhelmed. They may lash out, shut down, or feel flooded with anxiety.

Before substances: Many people with addiction have overactive stress responses due to trauma, anxiety, or chronic adversity.

During use: Substances numb the stress temporarily.

After addiction sets in: The brain becomes more sensitive to stress. Without the substance, the world feels overwhelming.

Recovery, then, isn't just about removing the substance. It's about regulating the stress system; helping the brain feel safe, grounded, and calm. Practices like surrender, prayer, mindfulness, and accountability (Steps 1–3 and 11) help soothe the amygdala and restore a sense of emotional stability.

The Social Brain
Disconnection and Isolation

We are wired to connect with others. Human beings need relationships to survive and thrive. That's not just psychology, it's neurobiology. The social brain includes networks that support empathy, bonding, and trust. It involves:

Oxytocin: A hormone often called the "bonding chemical," which helps us feel safe and connected.

Mirror neurons: Brain cells that allow us to feel what others are feeling, essential for empathy and understanding.

Anterior cingulate cortex: A part of the brain that helps us track social pain (like rejection or shame) as if it were physical pain.

Many people who develop substance use disorders have experienced social disconnection: loneliness, neglect, abuse, rejection. The addiction develops in a context of social pain. Often, the substance becomes a stand-in for what relationships once provided: comfort, relief, belonging, recognition from others or safe haven in a seemingly hostile or complex world.

Over time, addiction deepens isolation. It replaces real community with chemical connection. And because shame often accompanies addiction, people begin to hide, lie, or withdraw, further silencing the social brain. One of the most healing aspects of recovery is the return to co-regulation, the process of regulating our emotions in the presence of others. When someone listens to us without judgment, hugs us, tells the truth alongside us, or simply sits with us in pain, it actually calms our nervous system. The 12 Steps rebuild this capacity for connection, especially in Steps 5, 9, and 12.

In neurobiological terms, love, trust, recognition and belonging are medicine.

Addiction Is a Learning Disorder, Too

Many researchers now describe addiction as a form of disordered learning. In the brain, learning means making associations: this leads to that; this keeps me safe; this gives me relief. In other words, the brain learns to associate the substance with relief, reward, or survival, and unlearns how to pursue those things in healthy ways.

When substances repeatedly provide fast relief or pleasure, the brain learns that the substance is the solution. This "learning" becomes etched into memory and behavior, even when it leads to harm. Recovery requires a process of unlearning and relearning, teaching the brain that there are healthier, more sustainable ways to meet its needs. The 12 Steps offer a form of structured re-learning:

- *Step 1 teaches surrender instead of control*
- *Step 4 teaches reflection instead of avoidance*
- *Step 9 teaches repair instead of shame*

- *Step 12 teaches purpose instead of self-centeredness*

This kind of change doesn't happen all at once. It takes practice, repetition and supportive, compassionate community. But just as the brain learned addiction, the 12 Steps offer a structure for deep re-learning, through honest reflection, emotional repair, spiritual practice, and service to others.

So, What Needs Healing?

Here's a summary of what addiction breaks, and what recovery must help rebuild:

Brain Function	Addiction Impact	Recovery Restoration
Reward & Motivation	Dopamine overload, numbed pleasure	Meaning, hope, healthy habits (Steps 2, 10,12)
Impulse Control	Weakened prefrontal cortex	Structure, honesty, accountability (Step 10)
Stress Regulation	Hyperactive fear centers	Surrender, trust, mindfulness, (Steps 1-3, 11)
Connection & Empathy	Shame, Isolation, Social Disconnection	Confession, service, community (Steps 5, 9, 12)

A Final Word
Science Meets Grace

If this chapter has sounded technical at times, take heart: this isn't about reducing the miracle of recovery to molecules and neurons. It's about understanding what addiction actually does, and why the path of recovery, including the 12 Steps, so often works.

When we realize that the 12 Steps engage the very systems addiction has damaged, we can stop seeing them as mysterious or arbitrary. They're not magic, but they do tap into deep patterns of human biology, behavior, and change. Understanding how the brain works doesn't eliminate the need for grace; it helps us recognize where grace is already working.

In the next chapter, we'll begin our journey through the Steps themselves, starting with Step 1, and the hard but liberating truth of powerlessness.

Part II
The 12 Steps and the Brain

Chapter 2
Step 1 – Powerlessness and the Prefrontal Cortex
Why Admitting Defeat Strengthens the Brain

We admitted we were powerless over alcohol,
that our lives had become unmanageable.
Step One of Alcoholics Anonymous

Step 1 is the threshold of recovery. For many people, it is also the hardest. It asks us to admit something we've been running from: that we've lost control. That we can't think, manage, or willpower our way out of addiction. That what once helped us cope is now destroying us, and we can't stop on our own.

Spiritually, Step 1 is about surrender. Psychologically, it's about honesty. But what does this step look like from the perspective of the brain? Surprisingly, admitting powerlessness may be one of the most neurologically empowering things a person can do. That's because the very act of letting go, of facing the truth without resistance, calms the parts of the brain that addiction has thrown into chaos. In particular, it begins to unburden the prefrontal cortex, the area responsible for decision-making, self-control, and long-term thinking.

To understand why Step 1 matters so much, not just spiritually but biologically, we need to take a closer look at what the prefrontal cortex does, and how addiction disrupts it.

The Prefrontal Cortex
Our Inner "CEO"

The prefrontal cortex (PFC) is located right behind the forehead, and it's often described as the brain's executive center. Think of it as the part of your brain that does the adulting; it plans, makes decisions, considers consequences, and tells you not to yell at your boss or eat that third donut at midnight.

Some of its key roles include:

- *Impulse control: helping you pause before acting*
- *Moral reasoning: helping you consider right and wrong*
- *Future planning: weighing short-term vs. long-term consequences*
- *Self-awareness: monitoring your own behavior, thoughts, and feelings*

In short, the prefrontal cortex is the "you" that you're trying to be when you're thinking clearly and acting wisely. But here's the problem: addiction weakens this part of the brain.

Addiction and the Breakdown of Control

Addictive substances hijack the brain's reward system (as we saw in Chapter 1), flooding it with dopamine and teaching the brain that using is the top priority. Over time, this process deactivates and disrupts the prefrontal cortex, making it harder to:

- *Resist urges*
- *Delay gratification*
- *Reflect on the consequences of using*
- *Follow through on decisions to quit*

When someone says, "I knew I shouldn't have done it, but I couldn't stop myself," that's not an excuse, it's a neurobiological reality. The part of the brain responsible for rational control has been compromised. This is why Step 1 is so powerful. It's not just a spiritual confession; it's a biological shift. When someone admits, "I can't control this," they stop forcing their damaged executive brain to do a job it can't do alone. They relieve it of an impossible burden.

In neurobiological terms, surrender lightens the cognitive load.

The Exhaustion of Denial

Trying to stay in control when you're losing it is mentally exhausting. Many people in active addiction live with intense internal conflict: part of the brain says, "I shouldn't do this," while another part screams, "Do it anyway!" This inner tug-of-war taxes the prefrontal cortex, which already has limited capacity. When constantly overloaded with guilt, justification, and internal negotiation, the system begins to shut down, leaving people reactive, impulsive, and emotionally volatile.

Step 1 provides neurological relief. By saying, "I am powerless," the brain no longer has to maintain the illusion of control. This shift frees up mental and emotional energy for other tasks, like learning, reflecting, and healing. It's counterintuitive, but true: admitting defeat can be a step toward regaining cognitive strength.

Self-Awareness and the First Step Toward Healing

The decision to stop denying and face the truth activates another important function of the prefrontal cortex: self-awareness. This includes the brain's ability to:

- *Monitor our thoughts and behaviors*
- *Recognize harmful patterns*

- *Reflect on our motives and emotions*

Step 1 strengthens these abilities by creating space for honest reflection. Rather than fighting for control, we begin to observe our behavior without defending it. We stop justifying and start noticing. This is the beginning of metacognition, a fancy term for "thinking about our thinking." It's a uniquely human ability, and it plays a major role in long-term recovery. The Steps nurture this awareness gently but steadily. Step 1 is where it starts.

Humility as a Brain-Builder

We don't often associate humility with brain science, but it plays a surprisingly important role in how the brain opens up to change. In the context of recovery, humility isn't about humiliation or low self-worth. It's about accurately seeing our limits, and being willing to learn, listen, and grow.

Neurobiologically, humility helps lower the brain's defenses, specifically in the emotional and threat-detection regions of the brain, including the limbic system. The limbic system is a set of structures deep in the brain that processes emotion, memory, and survival instincts. One key player here is the amygdala, the part of the brain that scans for danger and triggers fight-or-flight responses when we feel threatened: physically, emotionally, or socially.

When we're caught in denial or self-justification, the brain is often responding to an unconscious threat: "If I admit this, I'll be judged." "If I'm wrong, I'll lose control." These thoughts activate the amygdala and keep us in a reactive state. We argue, rationalize, blame, or shut down. The emotional brain takes over, and the prefrontal cortex, the reasoning, reflecting part, gets pushed into the background.

But humility does something powerful: it signals to the brain that it's safe to let go of defenses. When we honestly say, "I was wrong," or "I don't have this figured out," we deactivate the alarm system. There's no threat to defend against, only reality to accept. In this way, humility softens the limbic response. It reduces emotional reactivity and invites the prefrontal cortex back online. That shift, from reactive to reflective, is essential in early recovery, and Step 1 creates the environment for it to begin.

In simpler terms: humility quiets the emotional storm so the thinking brain can reengage. This sets the stage for healing. Once the brain is no longer stuck in survival mode, it can begin to rewire - also referred to as neuroplasticity. It can start forming new pathways; new habits, new insights, new ways of relating to others and to oneself. That's what we mean when we say humility is a brain-builder. It doesn't just open the heart; it opens the mind.

Spiritual Surrender Meets Brain Science

From a spiritual perspective, Step 1 is about admitting that human willpower alone can't overcome addiction. That might sound disempowering, but in neurobiological terms, it's an accurate assessment of what the addicted brain can and cannot do.

Step 1 allows us to shift the work from a damaged system (the impaired prefrontal cortex) to a larger system; one that includes community, structure, support, and (for many) a Higher Power. By doing so, we reduce mental overload, increase honesty, and begin the slow process of repairing cognitive function. This is what makes Step 1 not just a confession, but a reset, for the mind and the nervous system.

Life Example
I Can't, But We Can

Take John, a 39-year-old in early recovery from alcohol and cocaine use. For years, John believed that if he just tried harder; if he read the right book, changed cities, switched jobs, he could get clean on his own. But every attempt ended the same way: he'd make it a few weeks, then relapse after a stressful event.

John's prefrontal cortex was doing what it could, but it was working against a hijacked reward system and a hypersensitive stress response. His willpower was outmatched. When John finally said the words, "I am powerless," it wasn't defeat, it was a beginning. The relief he felt wasn't just emotional, it was neurological. He no longer had to fight his brain alone. In admitting he couldn't do it by himself, he began building the neural capacity to learn a new way of living.

The Paradox of Step 1

Here's the paradox: admitting you've lost control is how you begin to regain it. Not by taking back the wheel through force, but by letting go of a strategy that never worked, and building a brain that can grow, adapt, and recover. By laying down the burden of control, Step 1 opens the door to something far better: the capacity to be transformed: biologically, emotionally, and spiritually.

Reflection Questions

Can you recall a moment when you admitted powerlessness? How did your body or mind respond?

What does surrender feel like to you: frightening, relieving, confusing?

In what areas of your life do you still try to "think" or "willpower" your way through?

How might your brain, and your recovery, benefit from letting go?

What are the key parts of the brain that are calmed by admitting powerlessness over our addiction?

Key Neurobiological Claims for Step 1

Addiction weakens the prefrontal cortex.

The brain's executive center, responsible for impulse control and long-term planning, becomes impaired, leaving decision-making dominated by craving and short-term reward.

The limbic system drives compulsive use.

When the prefrontal cortex is weakened, the amygdala and reward circuits dominate, making drug- or alcohol-seeking feel urgent and automatic.

Admitting powerlessness helps recalibrate brain systems.

Naming the loss of control interrupts denial and begins shifting balance away from overactive limbic drive toward prefrontal regulation.

Denial is partly neurobiological.

The brain often minimizes or rationalizes harmful patterns to reduce inner conflict; a process tied to distorted prefrontal and limbic interactions.

Moving from Step 1 to Step 2 with Lowered Defensiveness and Increasing Integration

Recognizing powerlessness calms inner conflict, paving the way for healthier cooperation between cognitive control, emotion regulation, and motivation systems. Next, we'll explore Step 2: Hope and the Dopamine System, and how belief in a power greater than ourselves reignites the brain’s ability to feel motivated, connected, and alive again.

Chapter 3
Step 2 – Hope and the Dopamine System
How Belief in Something Greater Rekindles Motivation and Meaning

Came to believe that a Power greater than ourselves could restore us to sanity.
Step Two of Alcoholics Anonymous

Step Two of Alcoholics Anonymous

If Step 1 is about surrendering control, Step 2 is about recovering hope. After admitting powerlessness, the next logical question is: If I can't fix this, who or what can? Step 2 invites us to consider the possibility, no matter how small, that healing is available, and that something greater than our own willpower might restore us to sanity.

That "Power greater than ourselves" has been interpreted in many ways: God, community, conscience, love, nature, the program itself. This step does not demand certainty; it invites openness. And from a neurobiological perspective, that shift from despair to possibility is transformative.

In this chapter, we'll explore how Step 2 interacts with the dopamine system, the brain's primary engine for motivation, desire, and meaning. Even the faintest flicker of hope, when nurtured and believed, can begin to restore this system after it has been numbed and hijacked by addiction.

Dopamine
The Fuel of Motivation

In Chapter 1, we introduced dopamine as a neurotransmitter, a chemical messenger that helps regulate the brain's reward and motivation system. But let's go a little deeper.

Dopamine isn't just about feeling good. It's about anticipating good. It's what drives us toward goals, gives us a sense of momentum, and helps us evaluate whether something is worth pursuing. In other words, dopamine is the fuel that gets us out of bed in the morning, not because we're happy, but because we believe something could happen that's worth doing.

When dopamine is flowing in a healthy way, people feel:

- *Curious and interested*
- *Motivated to pursue meaningful goals*
- *Able to anticipate reward or satisfaction*
- *Emotionally engaged in relationships and activities*

But in addiction, the dopamine system gets rewired. Substances like alcohol, heroin, cocaine, and methamphetamine flood the brain with dopamine, far beyond natural levels. Over time, the brain adapts by producing *less* dopamine and their receptors naturally. This leaves the brain less sensitive to the stimulation of normal pleasure and more "dependent" on substances for motivation.

This condition is called dopamine down-regulation, and it may lead to:

- *Emotional numbness or "flatness"*
- *A sense of pointlessness or futility*
- *Low energy, low motivation*
- *Cravings for intense stimulation (like drugs or risky behaviors)*

People in early recovery often describe this as a feeling of being dead inside. Nothing seems worth doing. Joy feels impossible. Life feels boring or hopeless. When it's not understood, this is the condition that reinforces the drive to keep drinking or using. This is not a failure of character, it's the result of a dopamine system that has been burned out and shut down.

How Step 2 Begins to Heal the System

When a person "comes to believe" that healing is possible, however vaguely defined, something begins to shift in the brain. That shift is hope, and hope is dopamine-rich; a driver of thoughts and feelings that can propel us toward a more positive future. Hope doesn't need to be perfect or fully formed. In fact, Step 2 doesn't say we must believe in a Higher Power, it says we came to believe. It acknowledges that belief is a process. And that process itself creates forward motion in the brain.

Here's what starts to happen:

- *The brain begins to anticipate that positive change might or could occur*
- *Dopamine levels begin to rise in response to this new expectancy*
- *Motivation becomes slightly more accessible*
- *Emotional numbness begins to thaw, even if just a little*

This doesn't happen overnight. But belief in the possibility of restoration; especially when supported by relationships, stories, and community, begins to reverse the neurochemical spiral of addiction. In this sense, Step 2 reactivates the dopamine system through the emotional and cognitive pathway of hope.

Belief and the Brain's "Meaning Network"

When people begin to believe in something beyond themselves, whether it's God, community, purpose, or simply the possibility of healing, they engage a powerful part of the brain called the default mode network, or DMN. Understanding this network gives us a deeper appreciation of why Step 2 is more than spiritual poetry; it's a neurological turning point.

What Is the Default Mode Network?

The default mode network (DMN) is a set of interconnected brain regions that becomes active when we're not focused on the outside world, when we're daydreaming, remembering the past, imagining the future, or reflecting on who we are. It's most active during moments of introspection, self-awareness, and internal narrative building. Spiritual or transcendent experiences, such as prayer, awe, meditation, or even deeply meaningful music, often activate this network.

In contrast, when we're doing a task, like solving a math problem or navigating traffic, other networks take over (such as the executive attention network or salience network). These "task-positive" networks are outwardly focused and shut down the DMN to keep us on track.

So why does the DMN matter for Step 2?

Because the DMN is where the brain begins to answer questions like:

- *Who am I?*
- *Why am I here?*
- *What is my life about?*
- *What do I really believe in?*
- *Can I imagine a future different than my past?*

These are exactly the kinds of questions that Step 2 invites us to ask.

How Step 2 Activates the DMN

Step 2 encourages people to consider that they are not alone, and that healing may be possible through a power greater than themselves. Whether that power is religious, relational, or philosophical, this step opens the door to meaning making.

As people begin to entertain these thoughts, the DMN becomes more active in a healthy way. This reflective activity:

- *Supports a stronger sense of self*
- *Enhances the ability to imagine a different future*
- *Improves emotional regulation and resilience*
- *Reduces anxiety, especially when paired with meditation or prayer*
- *Allows for more compassionate self-evaluation rather than shame-driven thinking*

In addiction, the DMN can become dysregulated, often reinforcing negative self-narratives ("I'm hopeless," "I'm broken," "I can't change"). But when people begin to believe in a story of restoration; a vision of sanity, healing, or divine care, the DMN begins to reorganize around new meaning.

This shift isn't just theoretical, it's functional. It lays down new neural pathways that shape how we interpret our past, how we perceive our present, and how we envision our future.

Meaning as a Motivational Force

The DMN doesn't act alone. When its reflections lead to a hopeful or purposeful outlook, it works with the dopamine system to reawaken motivation. In other words: The DMN says: "This life could be meaningful" and the dopamine system responds: "Then it's worth pursuing."

That's why Step 2 doesn't just feel like a turning point emotionally, it actually is a turning point neurologically. It re-engages the brain's deep wiring for identity, future orientation, and spiritual openness. That combination creates a foundation strong enough to carry us through the hard work of the steps ahead.

From Fatalism to Possibility

Addiction is often marked by fatalism, the belief that nothing can change, that one is destined to fail or suffer. This outlook is deeply demotivating, and it keeps the dopamine system in a depressed state. Step 2 counters that fatalism by creating space for possibility, even if it's just the possibility that other people have changed, and maybe we can too.

Imagine someone saying, "I don't know what I believe, but something is working for those people in that meeting." That thought alone, especially when followed by action, can help rekindle the brain's motivational circuits. Observation and openness are the first signs that the brain is waking up from the freeze of despair.

Life Example
Hope in Practice

Ana, a 28-year-old heroin addict, had relapsed seven times. She was deeply skeptical of anything spiritual and wanted nothing to do with a "Higher Power." But in a detox unit, she met a nurse who was also in recovery. The nurse told her: "You don't have to believe in God right now. Just believe that I believe. Borrow that until you believe it for yourself."

That moment became a turning point, not because Ana had a sudden conversion, but because she had a dopamine spark. She saw someone who had made it through. She felt the faintest sense of, Maybe I can, too. That thought gave her enough motivation to attend her next meeting, where she heard more stories of hope. The system began to rebuild. Ana's belief wasn't theological, it was neurochemical. And it was enough to start the next step forward.

What If I Don't Believe?

One of the great strengths of Step 2 is its openness. It doesn't demand certainty. You don't have to define your Higher Power, or even be sure that one exists. What matters is the willingness to consider that there is a path back to sanity and wholeness, and that the path doesn't rely on your own broken willpower and diminished motivation alone.

This willingness alone begins to shift brain activity away from despair and helplessness toward learning, openness, and growth. And it's here that spiritual recovery and brain science deeply agree: hope heals.

Reflection Questions

What does the phrase "Power greater than ourselves" mean to you at this stage in recovery?

Can you identify a moment when you began to believe that change was possible?

How does hope feel in your body or mind? What does it make you want to do?

Who or what helps you reconnect with a sense of meaning and forward motion?

Can you describe how music, prayer, meditation or nature help "rewire" the addicted brain?

Can you describe how the default mode network can lead to greater meaning and motivation?

Key Neurobiological Claims for Step 2

Dopamine is the brain's primary motivational signal, not just a pleasure chemical.

It drives anticipation and goal-directed behavior, making hope and belief powerful instigators for change.

Addiction blunts the dopamine system through overuse and downregulation.

This leaves individuals emotionally flat, unmotivated, and unable to find joy in normal life.

Hope and belief in recovery help reactivate the dopamine system.

Even tentative, borrowed or belief observed in others can spark renewed motivation and emotional engagement.

Spiritual reflection and openness engage the brain's meaning network (default mode network).

This supports long-term thinking, personal values, positive introspection and emotional regulation.

The experience of meaning and purpose improves mood and motivation.

This makes Step 2 a biologically relevant practice for reorienting the brain toward health, not just a spiritual or philosophical concept.

In the next chapter, we'll look at Step 3: Surrender and Stress Regulation and explore how letting go of control helps calm the brain's fear systems and begins to restore peace at the biological level.

Chapter 4
Step 3 – Surrender and Stress Regulation
How Letting Go Helps Calm the Fear-Driven Brain

Made a decision to turn our will and our lives over to the care of God as we understood Him.
Step Three of Alcoholics Anonymous

If Step 1 is about admitting we've lost control, and Step 2 is about believing that help is possible, then Step 3 is about making a choice: to stop fighting alone and to entrust our lives to something greater than ourselves. In spiritual terms, this is surrender. Not surrender in the military sense, where we admit defeat to an enemy, but surrender in the healing sense: letting go of the illusion that we can control everything, and placing ourselves in the care of something wiser, steadier, and more trustworthy than our ego or fear.

From a neurological perspective, this kind of surrender has profound effects. It lowers stress, quiets fear circuits in the brain, and begins to rewire the body's internal alarm system. In particular, Step 3 may help calm the amygdala, regulate the hypothalamic-pituitary-adrenal (HPA) axis, and build the conditions for emotional and physiological safety. In short: surrender changes your brain by turning off survival mode.

Stress and the Hijacked Brain

Many people with substance use disorders live in a chronically activated stress state. Even before drugs or alcohol enter the picture, the brain and body may be operating on high alert due to trauma, neglect, anxiety, or fear of failure.

Once addiction takes hold, stress becomes both a trigger and a consequence of substance use:

- *Stress increases cravings*
- *Cravings increase anxiety*
- *Relapse leads to shame and further stress*
- *This vicious loop keeps the limbic system, especially the amygdala, constantly activated.*

The amygdala is the brain's fear center. As mentioned, its job is to detect threats and initiate fight, flight, or freeze responses. In early humans, this was vital for survival. But in modern life, especially in addiction, the amygdala can become hypersensitive, reacting not just to physical danger but to emotions like guilt, grief, or uncertainty.

When the amygdala is firing, the prefrontal cortex (our reasoning and regulation center) can be suppressed. That's why stress can make people reactive, impulsive, or overwhelmed. But here's the good news: when we feel safe, the amygdala quiets down, and the prefrontal cortex re-engages. That's where Step 3 comes in.

The Neurobiology of Surrender

Surrender, in the Step 3 sense, is not giving up; it's letting go of control that isn't working. Spiritually, it means placing trust in something beyond ourselves. Biologically, it means interrupting the stress circuits that keep us stuck in survival mode.

To understand this, let's take a closer look at three key systems involved in the stress response:

The Amygdala
The Brain's Alarm System

The amygdala, that little, almond-shaped structure deep in the brain's limbic system has the primary task of detecting danger, whether physical, emotional, or social. When it senses a threat, it sends out an alarm signal that mobilizes the body to react quickly (fight, flight, or freeze).

In active addiction, or even in early sobriety, the amygdala often becomes hyperactive, meaning it's constantly on high alert. It doesn't just react to real dangers; it also responds to fear of failure, shame, uncertainty, or emotional discomfort. This chronic overactivation can lead to:

- *Panic attacks or irritability*
- *Emotional impulsivity*
- *Trouble calming down*

Step 3 helps quiet the amygdala by introducing the idea that we don't have to manage everything alone. When we hand things over: to God, to the group, to a Higher Power as we understand it, the brain interprets this as relief. The sense of being held or supported helps deactivate the alarm system. We have a "defense" system backing us up.

The HPA Axis
The Hormonal Stress Circuit

The hypothalamic-pituitary-adrenal (HPA) axis is the body's long-range stress response system. It connects the brain to the adrenal glands, which produce cortisol, the body's primary stress hormone. When the amygdala perceives a threat, it activates the HPA axis. This leads to:

- *Elevated cortisol levels*

- *Increased heart rate and blood pressure*
- *Suppressed digestion and immune function*
- *Heightened anxiety and emotional sensitivity*

Under normal conditions, moderate activation of the HPA axis can be helpful. But in addiction and chronic stress, the HPA axis is constantly active, which contributes to emotional exhaustion, poor sleep, and increased vulnerability to relapse. When we make a decision to "turn our will and our lives over," we interrupt this loop. Trusting in something greater, even temporarily, sends a neurological signal that says: the threat has passed, we're not alone, we can rest now. This allows the HPA axis to downregulate, reducing cortisol and shifting the body back toward balance.

The Parasympathetic Nervous System
The Body's Calming Circuit

The parasympathetic nervous system (PNS) is one half of the autonomic nervous system, the part that controls things like breathing, heart rate, and digestion without us thinking about it. The other half is the sympathetic nervous system, which revs us up in times of stress or danger. Think of the sympathetic system as the gas pedal, and the parasympathetic system as the brake.

When we're living in fear and control, the gas pedal is always pressed. We're tense, reactive, and worn out. But when we practice surrender, the PNS is activated, especially through deep breathing, prayer, reflection, or emotionally safe community and relationships. This leads to:

- *Slower heart rate*
- *Deeper breathing*
- *Improved digestion and sleep*
- *A feeling of calm and groundedness*

In other words, Step 3 gives the nervous system permission to exhale.

Cortisol Is Critical for Understanding Step 3

Cortisol plays a central role in the power of surrender. The stress hormone is released through the HPA axis when the brain perceives threat, and this mobilizes energy and sharpens attention for survival. It prepares the body and brain for action. In addiction this system becomes chronically activated, keeping cortisol levels elevated and reinforcing fear-based learning and trigger sensitivity in the amygdala. This may in turn weaken prefrontal self-regulation and disrupt the hippocampus's ability to provide emotional context.

Over time, the nervous system remains stuck in a state of vigilance, making control feel necessary, and the thought of "letting go" feel dangerous. Step 3 has the potential to interrupt this cycle because a conscious decision to surrender signals safety to the brain. The goal is to reduce HPA-axis activation, which can help cortisol levels gradually settle. When cortisol decreases, the brain becomes more capable of trust, reflection, and emotional regulation, creating the biological conditions for recovery to deepen.

Taken in this context, prayer and surrender are not passive spiritual ideas. They can also be understood and approached as active neurobiological interventions.

Why This Matters for Recovery

When someone takes Step 3, especially if they do it honestly and repeatedly, they begin to shift out of the chronic stress state that addiction reinforces. The amygdala softens, the HPA axis quiets, cortisol decreases, and the parasympathetic system comes online.

This creates the internal conditions for reflection, healing, and connection. Without this shift, the brain remains in "survival mode," and higher-order recovery work like inventory, amends, or service becomes difficult or even impossible. By helping to create the conditions for a biological reset, surrender has its own kind of power.

The Power of Trust

Whether you believe in God, the universe, your higher self, or simply the process of recovery, Step 3 works because trust calms the brain.

When we believe that we are not alone; that someone or something has our back and is helping to carry the burden, the brain interprets this as a signal of safety. That perception reduces hypervigilance and allows deeper emotional processing.

Research shows that people who feel securely connected to others or to a spiritual source experience:

- *Lower baseline cortisol (the stress hormone)*
- *Reduced anxiety*
- *Improved immune function*
- *Greater capacity for emotional resilience and delayed gratification*

In other words, trust isn't just comforting, it's regulating.

This is why Step 3 often becomes a turning point in early recovery. People stop white-knuckling sobriety and start participating in something larger. And in doing so, they begin to settle: emotionally, physically, and neurologically.

Life Example
The Weight of Control

Kevin, a 45-year-old father of two, had been sober for 30 days but was constantly on edge. He worried about everything, his job, his kids, his past, and whether he'd relapse again. He was trying to "stay in control" by micromanaging every detail of his life.

His sponsor asked him, "How's that working for you?" Kevin admitted he was exhausted. That night, he got on his knees, not because he fully believed in God, but because he needed to stop holding everything together by himself.

He didn't hear a voice. He didn't see a light. But he slept through the night for the first time in weeks. His shoulders dropped. His chest loosened. The next morning, the world felt slightly more manageable. What changed wasn't just spiritual, it was neurological. Kevin gave his overworked stress system permission to rest.

Surrender and Emotional Regulation

Recovery requires the ability to feel hard emotions without being overwhelmed by them. This is called emotional regulation, and it depends on a well-functioning feedback loop between the limbic system (emotion center) and the prefrontal cortex (logic and control center).

When the brain is in fight-or-flight mode, emotional regulation becomes almost impossible. But surrender, and the sense of trust it brings, interrupts this cycle. It gives the nervous system space to:

- *Pause*
- *Reflect*
- *Breathe*
- *Choose*

Step 3 helps people begin building this capacity. It's not about perfect serenity. It's about moving from emotional chaos to manageable discomfort, a necessary condition for long-term growth.

A Gentle Invitation

It's important to note that Step 3 doesn't require you to have it all figured out. You don't need theological clarity or unwavering faith. The Step simply asks: Are you willing to stop relying solely on yourself, and trust the process of recovery, however you understand it?

That small decision can begin rewiring the brain's most deeply rooted survival responses. It doesn't mean you'll never feel fear or pain again. It means you'll stop facing it alone, and your brain will know the difference.

Reflection Questions

What does the word "surrender" mean to you in the context of recovery?

Can you recall a time when letting go brought a sense of relief or clarity?

What situations or feelings activate your internal stress alarm? How do you respond?

What would it mean, practically and emotionally, to turn your will and life over to the care of something greater?

Have you experienced your amygdala creating a fear response that overwhelmed your prefrontal cortex so that you couldn't think clearly?

What brain region can create so much fear that it compromises your ability to think clearly or make rational decisions?

Key Neurobiological Claims for Step 3

Addiction keeps the brain in chronic stress activation.

The amygdala and HPA axis are often overactive in addicted individuals, leading to increased fear, impulsivity, and emotional volatility.

Surrender reduces stress-related brain activity.

The decision to relinquish control helps deactivate the amygdala, lower cortisol levels, and engage calming parasympathetic responses.

Trust, whether spiritual or relational, creates a neurological sense of safety.

This reduces emotional reactivity and enables the prefrontal cortex to reengage for decision-making and emotional regulation.

Emotional regulation depends on calming the limbic system.

Step 3 facilitates this by relieving the brain of the exhausting burden of control and replacing it with a felt sense of support.

Letting go is not weakness, it's a regulatory strategy.

Neurobiologically, surrender strengthens rather than weakens the individual by stabilizing core emotional circuits.

In the next chapter, we'll turn to Step 4: Inventory and Memory Reconsolidation and explore how courageously examining our past helps the brain process stored emotional experiences, laying the groundwork for clarity, healing, and long-term change.

Chapter 5
Step 4 – Inventory and Memory Reconsolidation
How Self-Reflection Helps the Brain Rewrite Emotional Memory

Made a searching and fearless moral inventory of ourselves.
Step Four of Alcoholics Anonymous

Step 4 often causes anxiety, even among those committed to recovery. The idea of making a "fearless moral inventory" can feel intimidating, overwhelming, or confusing. Why would we want to revisit the worst moments of our lives, catalog our failures, or open up old wounds?

But from a neurobiological perspective, this step is not about shame, it's about integration. Step 4 invites us to revisit emotionally charged memories in a safe, structured way so the brain can update them, reprocess them, and liberate us from their unconscious control. This process, known in neuroscience as memory reconsolidation, is one of the brain's most powerful healing mechanisms. It allows us to change how we relate to our past, and in doing so, alters how we behave in the present and future. In other words, Step 4 is not about beating ourselves up. It's about giving the brain a second chance to process pain.

How the Brain Stores Emotional Memory

When we experience something emotionally intense, especially if it's threatening or traumatic, the brain stores that memory in a vivid, often fragmented form. This involves key structures in the limbic system, including:

- *The amygdala, which tags the memory with emotional intensity (especially fear, anger, or shame)*
- *The hippocampus, which organizes the memory in time and space*
- *The prefrontal cortex, which ideally helps regulate and make meaning of the experience*

But in addiction, these systems don't always work well together. High emotional arousal can lead to memories that are deeply encoded but poorly integrated. As a result past experiences, especially painful ones, can remain stuck and unresolved in the nervous system, triggering shame, resentment, or fear long after the events are over.

These unprocessed memories fuel self-destructive patterns, negative beliefs ("I'm worthless," "I always fail," "I am toxic," "I can't be trusted."), and unconscious triggers that drive relapse. Step 4 begins the process of breaking that loop.

What Is Memory Reconsolidation?

Memory reconsolidation is the brain's natural ability to update old emotional memories; to revise the way they're stored based on new experiences, insights, or context. This process allows the brain to reduce the emotional charge around painful memories and even "rewrite" what they mean to us over time. In essence, the memory becomes contextualized and "unstuck."

Here's how it works in simple terms:

A memory is reactivated.

When we recall a past experience, especially one tied to strong emotions, the brain pulls that memory out of long-term storage. In this moment, the memory becomes unstable and changeable, like a file opened on a computer.

The brain enters a "plastic" state.

Once the memory is active, the brain enters a state of neuroplasticity, meaning it becomes more flexible and open to change. But for reconsolidation to occur, something new must be introduced; a new piece of information, a new emotion, or a new interpretation.

New meaning is added to the memory.

If we revisit a memory while feeling safe, supported, or more self-aware, the brain begins to associate the memory with those new emotional conditions. For example:

- *If we remember a childhood humiliation while feeling compassion for our younger self, that compassion gets layered into the memory.*
- *If we recall a painful loss while grounded in present-day strength, the brain re-anchors the experience in resilience rather than helplessness.*
- *If we name a fear while feeling heard by someone else, the memory no longer lives in isolation, it's been witnessed and integrated.*

The memory is re-stored in an updated form.

After this new information is added, the brain saves the memory again, but differently. It may still exist, but it carries less emotional weight. The memory is now part of a coherent story, not a raw wound.

Why This Matters for Step 4

This is exactly what Step 4 invites: the reactivation of emotional memories (through honest inventory), the introduction of new emotional context (through safety, structure, and intention), and the eventual re-storage of those memories in a more integrated, less reactive way.

That's why many people say that after doing a thorough Fourth Step, they feel lighter or clearer, even before Step 5. They're not forgetting the past. They're healing their relationship to it. They are adding wisdom and perspective to the memory. This is the heart of memory reconsolidation: it doesn't erase the story. It reframes it, giving the brain a new way to carry the past without being ruled by it. Said another way, It's not just confession. It's healthy neurobiological editing.

Facing What We've Avoided

Addiction thrives on avoidance. Many people use substances to escape pain, regret, or unresolved emotions. Over time, avoidance becomes automatic; it's not just about dodging hard feelings, it's about not even knowing what we feel anymore.

Step 4 begins to reverse this by gently bringing hidden patterns into the light. And while this can feel uncomfortable at first, it re-engages the prefrontal cortex and strengthens emotional regulation. When we examine our past, we activate the hippocampus (memory organization) and the ventromedial prefrontal cortex, which helps us process emotion and moral reasoning. These areas begin to "talk" to the amygdala, calming its response and reducing the emotional charge around the memory.

This means we can finally say things like:

- *"Yes, that happened, but it doesn't control me anymore."*
- *"That was painful, but I'm not stuck there."*
- *"I can understand this part of myself without judgment."*

The Power of Writing and Structure

The creation of our moral inventory isn't just free-form journaling. The process typically involves columns or categories, beginning with our liabilities:

- *Who or what we resent*
- *Our unexamined fears*
- *Examples of selfishness or false pride*
- *Critical moments of dishonesty*
- *Experiences of jealousy and envy*
- *Times of laziness or procrastination*
- *Specific harms we have committed*

Next, a list of assets that we strive for help to balance our liabilities and provide positive qualities to strive for, including:

- *Forgiveness*
- *Courage*
- *Humility*
- *Honesty*
- *Contentment*
- *Action*
- *Change*

This structure isn't just moral, it's cognitive. It helps the brain engage with painful material without being flooded or lost in chaos.

Why does writing help? It slows down emotional reactivity. It allows thoughts and feelings to be externalized and seen more clearly. It recruits the left hemisphere of the brain (language and logic), which helps regulate the more emotional right hemisphere. In other words, writing gives shape to what was previously amorphous. And shape is what the brain needs to process and move forward.

Life Example
Telling the Truth on Paper

Sasha, a 32-year-old in early recovery, feared Step 4 because she thought it meant dredging up everything she hated about herself. But with encouragement from her sponsor, she approached it as a curious investigation, not a punishment. She started by listing her resentments and was surprised to find that behind each one was fear: fear of being abandoned, misunderstood, or powerless. That realization changed the way she saw her past. She wasn't just angry; she was hurt and scared.

By the time Sasha finished her inventory, her shame had softened. She still felt grief, but it was clearer, less tangled. She later said, "I didn't know you could look at your worst moments and not want to disappear." That's the power of structured self-reflection. That's the brain beginning to heal.

From Shame to Integration

Shame is often described as "the feeling that we are bad." It disconnects us from ourselves and others. When shame is unresolved, it can keep memories stuck in emotional quicksand.

Step 4, especially when followed by Step 5 (sharing the inventory), can begin to shift shame into integration, a sense that our story has shape, coherence, and value, even in its darkest moments. Neurobiologically, this integration process strengthens the connections between memory, meaning, and emotional regulation. It allows us to carry our past without being crushed by it.

Reflection Questions

What feelings come up when you think about making an inventory of your past?

Have you noticed any recurring emotional or behavioral patterns in your life? Where might they come from?

Is there a memory or story that still feels "unfinished" in your body or mind?

What would it mean to examine your past not to condemn it, but to understand it?

How does the act of writing an inventory help connect the logical and emotional hemispheres of our brain?

Key Neurobiological Claims for Step 4

Emotionally charged memories are often stored in fragmented or unintegrated ways.

This makes them reactive and influential long after the event has passed.

Memory reconsolidation allows the brain to update emotional memories when they are revisited in a new, safer context.

This process helps reduce the emotional charge and create meaning.

Writing down difficult experiences engages cognitive and language centers in the brain.

This helps regulate emotional intensity and supports integration of left- and right-brain processing.

Structured reflection activates the hippocampus and prefrontal cortex.

These areas support emotional regulation, memory organization, and moral reasoning, reducing the grip of shame and impulsivity.

Step 4 is not a moral burden but a neurological opportunity.

It allows the brain to edit old stories, dissolve toxic narratives, and create space for clarity and healing.

In the next chapter, we'll explore Step 5: Telling the Truth and Social Neuroscience and discover how sharing our inventory with another person not only reduces shame but also activates relational circuits in the brain that are essential for healing, connection, and long-term recovery.

Chapter 6
Step 5 – Telling the Truth and Social Neuroscience
How Sharing Our Story Helps the Brain Heal Through Connection

Admitted to God, to ourselves, and to another human being the exact nature of our wrongs.
Step Five of Alcoholics Anonymous

Step 5 is often described as the most vulnerable moment in the early recovery journey. After making a searching and fearless inventory in Step 4, this step asks us to go a step further: to speak it aloud, to another person. This act of confession can feel intimidating, if not outright terrifying. It means being seen. It means risking judgment. But time and again, people who take this step report feeling relief, release, and a deeper sense of connection than they expected.

Why does this happen? Because telling the truth in the presence of another human being is not just emotionally meaningful, it's biologically healing. When done in a safe and supportive environment, Step 5 activates the social brain, reduces the toxic effects of shame, and engages neurological systems involved in trust, empathy, and co-regulation. In short: sharing our story helps rewire our brain for social connection and self-acceptance.

The Power of Being Seen and Heard

As human beings, we are wired for connection. From infancy onward, our brains develop in response to relationships, especially relationships that offer attunement, where another person mirrors, understands, and responds to our inner world. They are attuned to our authentic self.

In neuroscience, this is called co-regulation, the ability of two nervous systems to influence and stabilize each other. When someone listens to us with empathy and without judgment, our brain actually begins to calm down. Heart rate slows. Breath deepens. Fear centers quiet. This kind of interpersonal safety is a powerful counterforce to the isolation and shame that addiction creates. Step 5 offers exactly that: a structured opportunity to be heard and held in a moment of honesty.

Shame, Secrecy, and the Brain

Unspoken shame is a kind of psychological poison. When we keep painful truths hidden, they don't just go away, they continue to activate the limbic system, especially the amygdala, which responds to social threats like judgment, exclusion, or humiliation.

In fact, research has shown that social pain, the fear of being rejected or disapproved of, overlaps with some of the same brain regions as physical pain. This means that secrecy and unspoken guilt can literally hurt, and chronic shame keeps the brain in a state of vigilance and self-protection.

Step 5 interrupts this cycle. When we speak our story aloud, especially in a nonjudgmental setting, we allow the brain to reappraise the situation. We may find that the story has been distorted by the echo chamber of our mind. And equally important, we learn, often for the first time, that we can be known and still loved. This begins to rewire how we perceive ourselves and how safe we feel in the presence of others.

Oxytocin, Mirror Neurons, and Trust

Several powerful brain systems come online when we practice honest, relational vulnerability, like in

Step 5. Two of the most important are oxytocin and mirror neurons, which work together to reduce fear, build empathy, and create a sense of connection that's crucial for healing.

Oxytocin
The Bonding Hormone

Oxytocin is often called the "love hormone" or "bonding hormone" because this neurotransmitter is released in the brain during moments of closeness, trust, acceptance and emotional safety. It plays a key role in forming deep human connections, from parent-infant bonding to adult friendships and spiritual intimacy. When you share something deeply personal with someone who listens with compassion and without judgment, your brain responds by releasing oxytocin.

This release of oxytocin:

- *Calms the amygdala (the brain's fear center)*
- *Lowers heart rate and stress levels*
- *Promotes a feeling of being safe and accepted*

In the context of Step 5, oxytocin helps transform what could be a shame-inducing experience into one of regulation and healing. Instead of reinforcing trauma or fear, the act of sharing creates new associations in the brain: "I can be vulnerable and still be okay. I can be honest and still belong." In other words, Step 5 provides space for authentic, regulated vulnerability, the kind that promotes healing instead of re-traumatization. This is critical in early recovery, where trust has often been broken, internally and externally. Step 5 helps begin the process of rebuilding it at a neurochemical level.

Mirror Neurons
Empathy in the Brain

Mirror neurons are a special type of brain cell that help us connect emotionally with other people. They fire both when we take an action and when we observe someone else doing it, especially when that action involves emotion, facial expression, or storytelling.

These neurons are part of what allows us to:

- *Sense what others are feeling*
- *"Feel with" someone instead of just "feel for" them*
- *Experience resonance in shared emotional moments*

In Step 5, when we speak honestly about our past and our listener responds with understanding or shared emotion, our mirror neuron systems sync up. This kind of neural alignment:

- *Creates a sense of emotional resonance*
- *Reduces the sense of isolation or shame*
- *Helps shift difficult emotions from being personal burdens to shared, processed experiences*

When we share honestly with someone who listens and resonates, who expresses understanding, this validation builds mutual understanding and emotional alignment. By bringing painful emotions into a safe, shared space the burden is both understood and dissipated by the act of empathy. This process supports what therapists and religious leaders sometimes call "witnessing," being deeply heard in a way that transforms pain into meaning and connection.

From Isolation to Integration

In addition to the molecular and cellular activity (oxytocin & mirror neurons) triggered by Step 5, this step also affects our brain in a more holistic way. Step 5 helps turn our internal world, once disorganized and hidden, into a shared story. This process of putting our experiences into language and saying them aloud is a powerful act of brain integration.

When we speak honestly:

- *The left brain (language and logic) helps make sense of what the right brain (emotion and memory) has been holding.*
- *The prefrontal cortex gets engaged, allowing us to reflect instead of reacting.*
- *The default mode network supports us in seeing ourselves within a broader story: past, present, and future.*

In short, we move from disconnection to coherence.

Life Example
The First Time Being Known

Carlos, a 40-year-old in recovery from meth and alcohol, had done his Fourth Step inventory but was dreading the Fifth. He feared being judged, especially by another man. But his sponsor reassured him: "Whatever you say, I've either done it, thought about doing it, or heard it before." As Carlos began reading his inventory, his hands shook. But his sponsor just nodded, listened, and occasionally said things like, "That must've been so hard," or "Thanks for trusting me with that." In essence, the sponsor was countering Carlos' anxiety with an empathy born of lived experience; a key strength of Step 5.

By the end, Carlos felt exhausted, but light. "It's like I'm not hiding from myself anymore," he said. "And I don't have to hide from anyone else either." In turn, the sponsor witnesses the emergence of a more honest, centered man. Recovery in real time. That sense of safe exposure, being known without rejection, is one of the most powerful neurological events in recovery.

The Brain Learns Through Relationship

A key insight from interpersonal neurobiology is this: the brain learns to regulate emotion and form a stable sense of self through relationship. Step 5 taps into this truth by pairing deep self-examination with interpersonal connection.

We begin to learn:

- *I can feel shame and still be accepted*
- *I can name my patterns and not be destroyed by them*
- *I can share my worst moments and still belong*

These are not just insights; they are new experiences that shape the brain. They build emotional flexibility, resilience, and relational safety. Step 5 may be difficult, but it's also one of the most neurologically transformative steps in the program.

Reflection Questions

What fears come up for you when you think about sharing your story with another person?

Have you ever experienced deep relief after telling the truth about your past? What happened?

What kind of listener helps you feel safe? What qualities matter most in that person?

How do you imagine it would feel to be both vulnerable and fully known, and yet accepted and even embraced?

Can you "feel" the effect of mirror neurons when talking to someone who is listening with empathy?

Key Neurobiological Claims for Step 5

Shame and secrecy activate the brain's social threat circuitry.

The amygdala and related areas treat fear of judgment or rejection as a threat to survival.

Confession in a safe, empathetic setting reduces stress and activates the social brain.

Co-regulation, empathy, and nonjudgmental presence help lower cortisol and increase emotional regulation.

Oxytocin is released during moments of trust and vulnerability.

This neurochemical supports bonding, safety, and openness, key outcomes of a well-supported Step 5.

Mirror neurons create shared understanding and emotional resonance.

When someone listens with empathy, both brains begin to align, promoting healing.

Verbalizing painful experiences integrates left and right brain processing.

This turns fragmented memory into coherent narrative and reduces emotional overload.

In the next chapter, we'll explore Step 6: Readiness and the Brain's Default Patterns, and consider how the willingness to change, however uncertain, activates self-awareness, challenges entrenched habits, and prepares the brain for deeper transformation.

Chapter 7
Step 6 – Readiness and the Brain's Default Patterns
How Willingness to Change Begins to Rewire Habitual Brain Activity

Were entirely ready to have God remove all these defects of character.
Step Six of Alcoholics Anonymous

At first glance, Step 6 might seem abstract; how do you measure "readiness"? Unlike previous steps that involved action (writing, speaking, deciding), this one seems to focus on an internal state: a willingness to be changed. But make no mistake: readiness is powerful. From a neurological standpoint, willingness opens the door to plasticity, the brain's ability to change itself. It marks the shift from self-examination (Step 4) and honest disclosure (Step 5) to sustained transformation. It's the moment when the brain begins moving away from survival patterns and toward growth.

Step 6 invites us to ask: Am I truly open to letting go of what no longer serves me? Even the parts of me that helped me cope, but are now holding me back? This step may not seem dramatic on the outside, but inside the brain, it's another turning point.

Habits, Defenses, and the Brain's Default Mode

To understand what Step 6 is doing neurologically, we need to look at the brain's default mode network (DMN), a system introduced in Chapter 3, and central to how we think about ourselves and our habits. The default mode network is a group of brain regions that becomes active when we're not focused on the outside world; when we're daydreaming, ruminating, or reflecting on ourselves. It's involved in:

- *Self-narratives (how we talk to ourselves)*
- *Predicting how others see us*
- *Replaying past experiences*
- *Anticipating future events*

While the DMN is important for identity and planning, it also has a downside: it can reinforce familiar patterns, especially ones rooted in shame, fear, or self-justification. When addiction or trauma has shaped our inner world, the default mode may replay unhelpful messages like:

- *"I'll never change"*
- *"This is just who I am"*
- *"I have to protect myself"*
- *"I don't deserve better"*

Step 6 challenges this circuitry by introducing a new possibility: what if I'm ready to be different? This doesn't force change, but it creates the conditions for it. The DMN can be reshaped through intention, repetition, and openness to new interpretations of the self. This step asks us to release the grip on dogmas that served our addiction.

Readiness and Neuroplasticity

Neuroplasticity is the brain's ability to change in response to new experiences. But plasticity doesn't happen randomly, it depends on:

- *Attention (what we focus on)*
- *Emotion (what we care about)*
- *Repetition (what we practice)*
- *Readiness (whether we're open to the new)*

Step 6 is about readiness.

When we say, I'm willing to let this go, we're telling the brain: this belief, this behavior, this emotional reaction; it's no longer necessary. That signal triggers an internal shift. Old neural pathways begin to weaken through non-use, and the brain starts scanning for new options. Even if we're not sure how to change, or what exactly character defect "removal" looks like, our willingness changes the rules of engagement in the brain. We are no longer protecting old defenses. We're no longer arguing for our limitations. We're open. And openness is the foundation of all learning.

The Paradox of Readiness

One of the beautiful, and challenging, realities of Step 6 is that it doesn't demand perfection. It doesn't even demand immediate transformation. It asks only for willingness. This is psychologically and neurologically sound. Forced change often triggers the amygdala, creating resistance, anxiety, or even shutdown. On the contrary, voluntary, self-directed change, especially when rooted in meaning and community, activates reward circuits and calms fear responses by reducing amygdala activity.

In other words, the brain is more likely to change when it feels safe and invited, not shamed or forced. Step 6 operates within that window: creating space for change without rushing it. It's the difference between persuasion and coercion.

Life Example
Letting Go of the Mask

Elena, a 29-year-old in recovery from opioid addiction, came into the program with a sharp edge. Her sarcasm and toughness had helped her survive but were also keeping people at arm's length. In Steps 4 and

5, she acknowledged her need to always appear strong. But in Step 6, she asked herself: *Am I ready to let go of this mask?*

She wasn't entirely sure. It had protected her for years. Yet she was willing to try. That readiness allowed her to start experimenting, letting herself be seen and vulnerable in small ways, lowering her guard and risking connection. Over time, her brain and the feedback from her relationships began to change.

But there is a key element here. Elena didn't force herself into vulnerability. And no one was coercing her to change. In fact, if Elena had felt external pressure to change, that same pressure could have triggered her old edgy, protective behavior. Instead, she allowed herself to be *ready for it*. On her own terms. And that made all the difference.

How Step 6 Prepares the Brain for Change

Let's summarize what's happening in the brain when we practice the posture of Step 6:

Step 6 Posture	Neurobiological Effect
Openness to change	Increases neuroplasticity and weakens rigid habit loop
Self-reflection	Activates default mode network and makes our ingrained patterns visible
Voluntary surrender	Calms amygdala and stress system
Intention to grow	Boosts dopamine and motivation for new behavior

We may not see visible results right away. But readiness is the seed. It tells the brain, we're no longer surviving, we're preparing to grow.

Reflection Questions

Are there patterns or traits you've relied on for survival that no longer serve you?

What does "entirely ready" mean to you? Can you be ready even if you're unsure what change will look like?

Are there areas of your life where you still resist help, insight, or healing?

How might your brain, and your life, respond if you made room for change, even without knowing all the answers?

Can you think back to a time when forced change activated your amygdala and caused fear, stress and resistance?

Key Neurobiological Claims for Step 6

The default mode network reinforces habitual self-beliefs and behaviors.

Without intervention, it tends to replay familiar, often negative narratives.

Readiness opens the door to neuroplasticity.

Being willing to change, even before knowing how, creates the mental conditions for rewiring.

Voluntary change reduces limbic resistance.

The amygdala is less likely to trigger fear or stress when change feels chosen rather than forced.

Openness shifts attention and motivation.

As the brain anticipates new outcomes, dopamine begins to support exploration and learning.

Step 6 is a neurological invitation, not a demand.

It prepares the emotional and cognitive systems to begin letting go of protective, but limiting, patterns.

In the next chapter, we'll explore Step 7: Humility and Ego Dissolution and examine how practicing humility changes the brain's sense of self, shifting focus away from self-centered survival and toward openness, growth, and relational connection.

Chapter 8
Step 7 – Humility and Ego Dissolution
How Letting Go of Ego Helps the Brain Open to Connection and Change

Humbly asked Him to remove our shortcomings.
Step Seven of Alcoholics Anonymous

In Step 6, we became entirely ready for change. Step 7 now invites us to take the next step, not by pushing harder, but by asking humbly. For many people, the word humility can be uncomfortable. It's often misunderstood as weakness, passivity, or humiliation. But in the context of recovery, and neuroscience, humility is something very different. It's a mental posture that allows us to step outside our ego, see ourselves clearly, and become teachable.

Spiritually, Step 7 is about surrendering pride. Psychologically, it's about becoming socially aware and receptive to growth. And neurologically, it's about softening the brain's fixed patterns of identity, reducing defensiveness, and cultivating the neuroplastic space needed for lasting change. Step 7 is where we stop asking, "How can I fix myself?" and begin asking, "What am I willing to release, and how can I be reshaped?" To humbly ask for our shortcomings to be removed is to release control over how we grow.

How the Ego Forms and Why It Needs Recalibration

When we talk about the ego in this context, we're not referring to arrogance or narcissism, but to the internal system that maintains our identity, boundaries, and self-story. It's the structure our brain uses to navigate and interpret the world, answer the question "Who am I?", and protect us from real or perceived threats. The ego begins forming in childhood. Through our early experiences, especially those involving approval, rejection, reward, or punishment, we develop core beliefs about:

- *Our worth and identity*
- *What we must do to be safe*
- *How others are likely to treat us*
- *What is acceptable or dangerous to feel*

These beliefs become neural patterns, networks of thought, emotion, and behavior that are repeated over time. When people live in addiction, many of these patterns become rigid and defensive. The ego says:

- *"If I show weakness, I'll be hurt"*
- *"If I don't control things, I'll be abandoned"*
- *"I'm not safe unless I'm in charge"*
- *"I'm not the problem, everyone else is"*

These beliefs are often automatic and are reinforced by neural patterns in the default mode network (DMN). As we discussed in earlier chapters, the DMN helps us construct a narrative self, but when left unchecked, it can keep us stuck in shame, pride, or blame. Step 7 challenges these patterns through humility, not self-deprecation, but right-sizing the ego so that we can stop defending who we were and start becoming who we could be.

Ego Dissolution and Spiritual Openness

Some neuroscientists have studied what happens in the brain during ego-dissolving experiences, such as deep meditation, awe, intense prayer, or psychedelic states. These experiences are often accompanied by:

- *A temporary loosening of the self/other boundary*
- *A sense of unity or interconnectedness*
- *Increased feelings of compassion and insight*
- *Long-term reductions in anxiety, depression, and addictive behaviors*

These shifts are associated with deactivation of the DMN and increased communication between brain networks that don't usually "talk" to each other. While Step 7 doesn't require a mystical experience, it does involve a similar kind of surrender: the release of the illusion that we are in total control of our growth. When we humbly ask for help, whether from God, community, or inner conscience, we shift from control to cooperation with something greater.

The Default Mode Network
Friend and Foe

The default mode network (DMN), the major brain system involved in self-awareness, introspection, and storytelling becomes active when we are thinking about ourselves, imagining the future, or revisiting the past.

In earlier chapters, we explored how the DMN plays a positive role:

- *It allows us to form meaning and personal identity*
- *It supports reflection and moral reasoning*
- *It enables spiritual imagination and hope*

But there's a shadow side. When overactive or dysregulated, the DMN can trap us in rigid narratives:

- *Rumination ("I can't let go of this thought")*
- *Shame loops ("Why am I like this?")*
- *Fantasy and avoidance ("I want to escape")*

For people living in addiction, the DMN often becomes dominated by critical, self-centered or fear-based thinking, reinforced by trauma or long-standing patterns of control. Step 7 helps recalibrate the DMN. By practicing humility, intentionally stepping back from the ego, we reduce over-identification with those rigid self-narratives. We shift the DMN from being a narrow loop of self-protection to a channel of self-transcendence.

Serotonin and Ego Flexibility

One of the key neurochemicals involved in this shift is serotonin, a neurotransmitter that helps regulate mood, emotion, and cognitive flexibility. Serotonin

- *Promotes emotional stability*
- *Helps regulate social behavior and impulse control*
- *Supports ego flexibility, our ability to shift perspective and de-center from ourselves*

Low serotonin can be associated with impulsivity, emotional reactivity, and rigid thinking; patterns often seen in active addiction. As serotonin levels begin to stabilize in recovery (often through abstinence, sleep, nutrition, and connection), the brain becomes more open to humility and change.

Some studies have even shown that increased serotonin activity in certain brain regions leads to a diminished sense of separateness; less "me against the world," and more openness to connection and shared meaning. In spiritual terms, this feels like humility. In neurological terms, it's reduced self-centeredness (ego salience) and increased relational awareness.

Humility in the Brain
The Role of TPJ and mPFC

Two key brain regions help support the experience of humility and the willingness to let go of ego-driven control:

Temporoparietal Junction (TPJ)

The TPJ is located at the intersection of the temporal and parietal lobes and is involved in perspective-taking and theory of mind, our ability to imagine the thoughts and feelings of others. In humility, the TPJ helps us recognize that:

- *Other people have valid points of view*
- *We are not the center of every story*
- *We can shift our stance and learn from others*

Step 7 activates this region when we begin to consider our limitations without shame and recognize the wisdom outside ourselves.

Medial Prefrontal Cortex (mPFC)

The mPFC is part of the brain's self-processing network. It helps us evaluate ourselves, reflect on our actions, and consider moral or social values. When functioning well, the mPFC allows us to:

- *Engage in healthy self-reflection*
- *Adjust our behavior based on feedback*
- *Connect personal growth to meaningful values*

Humility strengthens these circuits, allowing us to reflect without spiraling into self-criticism or defensive pride. It supports a balanced view of self, rooted in curiosity and growth rather than fear and control.

Life Example
Letting Go of Self-Fixing

Marcus, a 36-year-old in recovery from alcohol and porn addiction, had always prided himself on being "a fixer." In his Fourth and Fifth Steps, he identified a pattern: when he felt shame, he tried to solve it by performing, overachieving, or isolating. In some cases, he turned to "fixing" other people's problems as a means to bolster his own ego and cover his low self-esteem. In Step 7, he finally asked, "What if I can't fix any of this by myself?"

He described praying with the words: "I don't know what to do with this anymore. Help me let go." He didn't feel fireworks, but he noticed something different afterward. He stopped trying to manage everyone's impression of him. He asked for help more often. And even though he was now focused on cleaning up his own side of the street, he felt less alone. Marcus hadn't removed his shortcomings through force. He had made space for change, and that changed everything.

From Rigid Self to Open Soul

Step 7 doesn't require us to erase our identity. It invites us to loosen our grip on the parts of ourselves that are rigid, defensive, or rooted in fear. Humility, then, isn't weakness, it's a recalibration of power. It's saying, "I can't change myself by ego alone, but I'm open to being changed."

Neurologically, this means:

- *Softening overactive or destructively self-centered DMN patterns*
- *Strengthening circuits for empathy, reflection, and relational awareness*

- *Stabilizing serotonin levels to support perspective and peace*
- *Making room for growth by releasing old ego strategies*

Humility is not thinking less of ourselves; it's thinking of ourselves less, and learning to trust something bigger than our own effort. It's a neurobiologically potent move.

Reflection Questions

Where do you notice ego-driven defenses still operating in your life?

What would humility look like in your recovery, not just as a value, but as a daily practice?

How has your internal narrative changed since beginning the steps?

Are there parts of your identity you're ready to loosen in order to grow?

Can you imagine the temporal and parietal lobes of your brain transitioning from self-centeredness to the thoughts and feelings of others?

Key Neurobiological Claims for Step 7

The ego is built from adaptive neural patterns of identity and protection.

In addiction, these patterns become rigid, self-focused, and defensive.

The default mode network (DMN) supports both self-awareness and rumination.

Step 7 helps recalibrate the DMN from rigid ego loops (rumination) that don't allow us to relax our self-image, toward flexible, reflective growth and evolution.

Serotonin promotes ego flexibility and emotional stability.

As serotonin stabilizes in recovery, it supports openness to change and reduction in self-centered reactivity.

Humility engages brain regions that support perspective-taking and moral awareness.

The temporoparietal junction (TPJ) and medial prefrontal cortex (mPFC) help us reflect wisely and interact with others maturely.

Step 7 fosters a neurological shift from self-protection to self-transcendence.

This opens the brain for deeper transformation, social healing, and emotional integration.

In the next chapter, we'll turn to Step 8: Empathy and Prefrontal Cortex Growth and see how making a list of those we've harmed exercises the brain's moral and relational networks, strengthens empathy circuits, and prepares the brain for the difficult but healing work of amends.

Chapter 9
Step 8 – Empathy and Prefrontal Cortex Growth
How Making Amends Begins with Building Empathy in the Brain

Made a list of all persons we had harmed,
and became willing to make amends to them all.
Step Eight of Alcoholics Anonymous

Step 8 begins the amends process, but it starts slowly and subtly: with a list and willingness. Before we speak to anyone or set things right, we are asked to pause and reflect on the harm we have caused, without excuses, avoidance, or blame. Spiritually, this step is about compassion and accountability, seeing our actions in a broader field of vision. Psychologically, it is about acknowledging the relational consequences of addiction. Neurobiologically, it is about building the capacity for empathy and moral reasoning, skills that are essential for sustained recovery.

Step 8 prepares us to restore relationships not only by making external repairs but by rewiring the brain for emotional maturity, perspective-taking, and social connection.

The Importance of Empathy in Recovery

Empathy is the ability to understand and share the feelings of another person. It means not just recognizing what someone else is going through, but also imagining how it might feel to be in their place. Empathy creates a bridge of connection, allowing compassion and support to flow between people. At its core, it is about tuning in to another's inner world with interest and openness. It involves:

- *Recognizing someone else's emotions*
- *Imagining their perspective*
- *Responding with appropriate care or accountability*

In this context, empathy is not about condemning yourself for the pain caused to others; it is thoughtful, mature connection. However, in addiction, empathy often becomes blunted:

- *The brain's reward system prioritizes the substance over relationships*
- *Chronic stress and shame lead to defensive, self-focused thinking*
- *Emotional pain narrows perspective to survival mode, making others' feelings secondary*

Step 8 begins to heal this by shifting our focus outward, asking us to consider, How did my actions affect others? This is a moral and neurological expansion from self-centered survival toward relational awareness.

The Role of the Prefrontal Cortex

The prefrontal cortex (PFC), as mentioned in Chapter 2, is responsible for:

- *Impulse control*
- *Long-term planning*
- *Perspective-taking*
- *Moral reasoning*

Active addiction weakens these circuits, making it difficult to consider consequences or the feelings of others. As sobriety progresses, the PFC begins to recover, but it needs exercise to grow stronger.

Step 8 provides that exercise. By listing those we've harmed and reflecting on how to make amends, we:

- *Practice delaying impulses (instead of rushing to justify or avoid)*
- *Engage moral reflection about right action*
- *Strengthen cognitive empathy, social perception and understanding how others feel*

This process supports prefrontal cortex growth, building the neurological pathways we need for healthy relationships and emotional sobriety.

Empathy in the Brain
TPJ and mPFC Revisited

Two brain areas highlighted in Step 7 continue to play a key role in Step 8:

Temporoparietal Junction (TPJ)

The TPJ is crucial for perspective-taking. It allows us to step outside ourselves and ask, How did my actions feel to them? Without TPJ engagement, we remain trapped in self-centered narratives. In Step 7, the TPJ helps provide us an inward perspective, to see ourselves more accurately and humbly. It enables us to reflect on our limitations without defensive ego distortion, supporting self-honesty and humility.

In Step 8, the TPJ shifts its function from an inward-focused perspective to outward-focused perspective-taking:

- *It helps us imagine how others experienced our behaviors, supporting cognitive empathy*
- *It allows us to mentally "stand in another's shoes," seeing beyond our intentions to the actual impact of our actions*
- *It prepares us to approach amends with understanding rather than just obligation*

The TPJ's role in the social cognition network means it works with the mPFC to help us:

- *Understand the emotions, thoughts, and needs of others*
- *Prepare to respond in a relationally attuned, respectful way*
- *Build the mental flexibility needed to see situations from multiple angles*

Thus, in Step 7, the TPJ helps us see ourselves accurately. In Step 8, the TPJ helps us see others accurately. This progression is critical in recovery, moving us from personal insight toward relational empathy and accountability, which sets the stage for action in Step 9.

Medial Prefrontal Cortex (mPFC)

The medial prefrontal cortex (mPFC) plays a central role in self-reflection, moral reasoning, and social cognition, the ability to understand ourselves in relation to others. Like many other brain regions, mPFC can dynamically shift its functions during different mental activities, such as in the different steps. This functional shift usually occurs through its engagement in different brain networks.

For example, in both step 7 and 8, mPFC can be highly involved, but with different functions in different brain networks. In Step 7, the mPFC is more involved in the default mode network (DMN), which helps us reflect on our limitations and need for growth, while supporting humility and openness. It is engaged as we examine ego-driven defenses and become willing to let them go, fostering self-awareness without self-condemnation.

In Step 8, the mPFC shifts from inward-focused reflection to relational reflection:

- *It helps us evaluate how our behaviors have impacted others*
- *It integrates personal accountability with social awareness, connecting our values to the real-life consequences of our actions*
- *It supports moral emotions such as guilt (in healthy, non-toxic forms), concern, and compassion*

This shifting is more engaged in the social cognition network, where it is working alongside the temporoparietal junction (TPJ) and other areas to:

- *Understand others' perspectives and emotions*
- *Predict how others may feel or respond*
- *Consider social norms and moral frameworks during decision-making*

In Step 8, this network is activated as we:

- *Imagine how our past actions felt for those we harmed*
- *Reflect on what making amends would mean for restoring trust*
- *Prepare to align our behavior with our values in our relationships*

Thus, in Step 7, the mPFC helps de-center the ego to allow growth. In Step 8, the mPFC helps extend that growth outward, supporting relational accountability and moral repair. By engaging the mPFC during this step, we strengthen the brain's capacity for empathy, ethical action, and relational healing; key foundations for making effective amends in Step 9.

The Dopamine System and Social Repair

Interestingly, acts of moral repair can engage the brain's dopamine system. While addiction hijacks this system for temporary intense pleasure, healthy connection and meaningful restitution can stimulate dopamine in a normal, sustainable level. The anticipation of restoring a relationship or reducing harm can generate a sense of purpose, motivation, and hope, helping rewire the reward system away from impulsivity toward meaningful goals.

Life Example
The List Before the Call

Tariq, a 33-year-old in recovery from alcohol and gambling addiction, felt overwhelmed when he began Step 8. "How can I face the people I hurt?" he asked. His sponsor reminded him: "You don't have to fix it all today. Just write the list." As Tariq wrote down each name: his mother, sister, former business partner, he felt waves of regret but also clarity. He saw patterns: lies told to avoid shame, debts left unpaid, promises broken.

By the time he finished, he felt more grounded. "For the first time, I saw the wreckage clearly," he said, "and I wanted to be the kind of person who cleans it up." Tariq's brain was changing. Perspective-taking circuits were activating, and his dopamine system began associating recovery with healthy human connection, not avoidance and escape.

Why This Step Matters Neurologically

Step 8 does not require us to act yet. It requires us to prepare internally. This preparation:

- *Activates empathy circuits (TPJ, mPFC)*
- *Strengthens moral reasoning (PFC)*
- *Shifts reward circuits toward meaningful goals (dopamine system)*
- *Reduces defensiveness, replacing it with compassionate accountability*

These are not small changes. They are the groundwork for becoming someone who can live in integrity.

Reflection Questions

Who have you harmed in your addiction, directly or indirectly?

What feelings arise when you think of making amends?

Will those feelings be accompanied by a healthy release of dopamine in the brain?

How might seeing your actions from others' perspectives help you grow?

Beyond addiction recovery, how can exercising and strengthening your prefrontal cortex improve your life?

Key Neurobiological Claims for Step 8

Addiction blunts empathy, prioritizing immediate reward over relationships.

Recovery reopens pathways for perspective-taking and moral awareness.

Step 8 exercises the prefrontal cortex.

Listing and reflecting build impulse control, planning, and relational thinking.

The TPJ supports perspective-taking, while the mPFC supports moral evaluation.

Both are strengthened when we reflect on those we have harmed.

Dopamine can support social repair.

Anticipating meaningful restitution and connection can engage motivation systems in healthy ways.

Preparation before action is neurological training.

Step 8 creates the brain conditions necessary for successful amends in Step 9.

In the next chapter, we will explore Step 9: Making Amends and Social Bonding, examining how taking action to repair harm not only restores relationships but also activates the brain's bonding and reward systems, deepening emotional sobriety and building a new way of living in community.

Chapter 10
Step 9 – Making Amends and Social Bonding
How Repairing Relationships Restores Brain Circuits for Trust and Bonding

Made direct amends to such people wherever possible, except when to do so would injure them or others.
Step Nine of Alcoholics Anonymous

In Step 9, we begin the work of repair. After naming those we've harmed (Step 8), we now seek to restore what was broken, through words, actions, or changed behavior. This step is both deeply personal and profoundly interpersonal. It can also be one of the most emotionally charged moments in the recovery journey.

Spiritually, Step 9 is an act of reconciliation. Psychologically, it is an act of accountability and healing. Neurobiologically, Step 9 engages brain systems responsible for social bonding, moral behavior, and emotional regulation, all of which are often disrupted by the broken relationships, trust deficit, anxiety, humiliation and isolation that result from prolonged addictive behavior. This step not only heals relationships, but it also helps repair the brain's ability to connect, trust, and thrive in community.

Why Relationships Are Central to Recovery

Humans are social beings. Our brains are wired to seek connection, belonging, and commitment. These needs are not just emotional, they are biological. From infancy to adulthood, relationships shape the brain's development. Addiction often damages this social wiring:

- *It isolates us from others*
- *It increases shame and defensiveness*
- *It erodes trust, honesty, and mutual care*

Even after physical sobriety is achieved, the relational wounds often remain. Step 9 addresses this by inviting us to take specific actions to rebuild interpersonal integrity and re-engage our social circuitry.

Social Bonding and the Brain
The Role of Oxytocin

One of the key neurochemicals involved in amends-making is oxytocin. As mentioned earlier, it is often called the "bonding hormone," and its main functions include:

- *Increases during acts of trust, apology, and mutual understanding*
- *Reduces activity in the amygdala, calming fear and defensiveness*
- *Enhances social bonding and emotional regulation*

When we make amends sincerely, and are met with compassion or forgiveness, oxytocin levels rise, helping both parties feel safer, more connected, and more open. Even when an amends conversation is difficult, the intention to repair can reduce internal stress and promote neurobiological integration. Just preparing to take responsibility can begin rebuilding social circuits damaged by addiction.

The Prefrontal Cortex in Action
Planning and Emotional Regulation

As we saw in Step 8, the prefrontal cortex (PFC) plays a key role in moral reasoning and empathy. In Step 9, it becomes even more active as we:

- *Evaluate when and how to approach each person on our list*
- *Consider what will be helpful or harmful in each context*
- *Practice emotional regulation during the actual conversation*

Making amends isn't a reflex, it requires executive functioning: planning, assessing risk, managing emotion, and aligning actions with values. The PFC coordinates all of this. Each time we act with integrity and intention, we strengthen the PFC's leadership role in the brain; analogous to exercising a muscle to make it stronger. This in turn reduces or helps counter-act the influence of impulsive or avoidant systems.

The Amygdala, Fear, and Courage

It's natural to feel fear when approaching someone we've hurt. The amygdala, as mentioned in previous chapters, is a key structure in the limbic system and responds to perceived threats; emotional or physical. Facing the possibility of rejection or conflict can activate this fear circuitry. However, research shows that taking meaningful action, especially when aligned with personal values, helps reduce amygdala reactivity over time. This is known as exposure with safety; when we face what we fear and discover that we can survive (and even grow from) an emotionally charged or ethically difficult encounter.

In conjunction with the amygdala-calming effects of oxytocin, making amends also rewires fear-based patterns and replaces them with relational courage.

The Neurobiology of Guilt and Resolution

Guilt is one of the most powerful, and potentially healing, emotions in recovery. Unlike shame, which tells us "I am bad," guilt says, "I did something that hurt someone, and I want to make it right." While shame often isolates, guilt can motivate reconnection and repair. Neurobiologically, guilt involves an interplay between emotional distress and moral reflection. When experienced in a healthy, regulated way, it leads to growth, not paralysis.

Brain Regions Involved in Guilt

Research suggests that guilt activates a network of brain regions responsible for:

- *Moral reasoning (medial prefrontal cortex, or mPFC)*
- *Perspective-taking (temporoparietal junction, or TPJ)*
- *Emotion regulation and social concern (anterior cingulate cortex and insula)*

This network overlaps significantly with the social cognition system engaged in Step 8 and Step 9. Guilt is essentially the emotional signal that tells us our behavior has conflicted with our values or harmed someone we care about. It's a moral warning.

Neurotransmitters That Support Guilt Resolution

When guilt is acknowledged and acted upon, especially through amends, it engages multiple neurochemical systems that support resolution and emotional healing and maturity:

GABA

The Calming Neurotransmitter

Gamma-aminobutyric acid (GABA) is the brain's primary inhibitory neurotransmitter. It acts like a brake on the nervous system, helping us calm down after stress or arousal. Making amends, especially when we're received with understanding, can increase GABA activity, leading to:

- *Reduced anxiety and emotional reactivity*
- *Lower amygdala activation*
- *A sense of internal relief and peace*

This is one reason why people often describe feeling a "weight lifted" after a meaningful amends.

Serotonin

The Regulator of Mood and Perspective

Serotonin helps regulate mood, self-reflection, and impulse control. Low serotonin levels are linked with rigid, shame-based thinking and a lack of perspective. When guilt is unresolved, the brain can spiral into rumination, which keeps serotonin low and the default mode network (DMN) overactive. But when guilt is acknowledged, felt, and followed by repair:

- *Serotonin activity increases*
- *Cognitive flexibility improves*
- *Emotional resilience strengthens*

This helps shift guilt from being a loop of regret into a pathway for growth.

Dopamine

The Reward of Making It Right

While dopamine is often associated with addictive behaviors, it also plays a crucial role in healthy motivation and satisfaction. When we make amends, especially when it leads to reconnection or closure, the brain may release dopamine, reinforcing the experience of:

- *Acting in alignment with one's values*

- *Rebuilding trust and relational faith*
- *Repairing identity and self-respect*

In this way, moral action becomes rewarding, helping the brain associate relational integrity with emotional safety and pleasure.

Step 9 as Guilt Integration

Step 9 doesn't eliminate guilt by denying it. It integrates guilt into the healing process:

- *By taking ownership, we acknowledge the emotional signal*
- *By making amends, we respond with action*
- *By reconnecting, we resolve the emotional tension in body and brain*

Rather than being overwhelmed by guilt, or numbing it as in active addiction, this step allows guilt to serve its true evolutionary purpose: to motivate repair and restore relational and communal balance.

Life Example
Choosing Repair Over Avoidance

When Rico began Step 9, he was afraid of facing his father. During his drinking years, he had lied, borrowed money he didn't repay, and disappeared for months at a time. Every attempt to think about making amends triggered the same bodily reaction: tight chest, racing heart, and an urge to escape. His sponsor walked him through the process slowly. "Don't start with the conversation," he said. "Start with the truth."

Rico wrote down exactly what he wanted to admit and take responsibility for. He practiced saying it aloud, feeling the mixture of fear and relief rising in his body. When he finally met with his father, he focused on staying grounded, breathing slowly, and staying present even as old guilt and shame surfaced.

His father didn't respond with anger. Instead, he sighed and said, "I just wanted my son back." Rico walked away from that meeting lighter than he expected—not because the relationship was instantly repaired, but because he had acted from courage rather than avoidance. His brain learned something new: guilt didn't have to isolate him anymore. Repair was possible, and his nervous system could survive honesty.

Step 9 as a Neuro-Social Reset

Step 9 works on multiple levels:

- *It repairs interpersonal trust and social bonds*
- *It rebuilds moral integrity*
- *It engages courage over avoidance*
- *It reduces limbic fear and activates prefrontal growth*
- *But most importantly, it gives the brain a new relational template: I can face what I've done. I can seek repair. I can reconnect.*

This is the opposite of shame. It is social re-integration, and for an addict, it's a process that offers immeasurable healing for the human body, brain and soul.

Reflection Questions

Which relationships from your Step 8 list feel most in need of healing?

What emotions arise when you imagine making amends?

How might taking responsibility help calm your nervous system and restore trust?

Are there people you cannot make direct amends to?

How might you honor the spirit of repair in other ways?

How do different neurotransmitters, the amygdala and the prefrontal cortex factor into the healing process of making amends?

Key Neurobiological Claims for Step 9

Oxytocin plays a central role in social bonding and repair.

It increases with trust-building behaviors like apology, listening, and reconciliation.

Resolving guilt engages other neurotransmitter systems for emotional regulation.

Neurotransmitters like GABA (calming), serotonin (mood regulation), and dopamine (reward) support emotional resolution and reinforce prosocial behavior.

The prefrontal cortex supports planning and emotional regulation.

Step 9 activates this system as we thoughtfully plan and engage in relational repair - the making of amends.

The amygdala generates fear in anticipation of conflict or rejection.

Facing these fears with courage and thoughtfulness helps reduce its dominance.

Making amends recalibrates the brain's relational circuitry.

It helps transition from shame and avoidance to courage, connection and accountability.

Step 9 is a powerful act of neural integration.

It aligns thought, emotion, and behavior around core values, promoting long-term change.

In the next chapter, we'll explore Step 10: Daily Inventory and the Brain's Self-Monitoring System, and see how regular reflection strengthens attention, emotional awareness, and long-term behavioral change, helping the brain stay grounded and resilient in recovery.

Chapter 11
Step 10 – Daily Inventory and the Brain's Self-Monitoring System
How Ongoing Self-Reflection Strengthens Emotional Regulation and Cognitive Control

Continued to take personal inventory and when we were wrong promptly admitted it.
Step Ten of Alcoholics Anonymous

Step 10 is about maintenance, keeping our recovery clean and current. While earlier steps address past harms and old patterns, Step 10 asks us to stay alert to the present moment: to notice when we've gone off course, take responsibility, and correct our path promptly.

Spiritually, Step 10 reflects living in integrity, awareness and humility on a daily basis. Psychologically, it is about practicing mindfulness, honesty, focus and accountability in real time. Neurobiologically, Step 10 strengthens the brain's self-monitoring and error-correction systems, helping us respond skillfully to life challenges rather than reacting impulsively to the kind of emotionally or socially charged situations that used to trip or trigger us.

This ongoing practice doesn't just prevent relapse, it builds neural resilience, keeping us anchored in recovery and connected to our values. And it becomes a solid foundation for healthy, long-lasting relationships. In other words, it's the behavior on which we build a better future.

The Brain's Self-Monitoring System

At the core of Step 10's process and power is self-monitoring, the brain's ability to:

- *Track our own actions and emotional states*
- *Detect when we've made an error or caused harm*
- *Adjust our behavior in response*

The anterior cingulate cortex (ACC) is central to this process. Located deep in the frontal part of the brain, the ACC:

- *Monitors for discrepancies between intention and action*
- *Triggers a "warning signal" when we act outside our values*
- *Works with the prefrontal cortex (PFC) to plan corrective action*

In addiction this monitoring system is often overridden by craving and impulse. In recovery, Step 10 re-trains the ACC–PFC loop to prioritize integrity and course correction. Think of the ACC as providing a kind of constructive internal feedback and criticism.

What Is the Anterior Cingulate Cortex (ACC)?

The anterior cingulate cortex is a C-shaped strip of brain just above the corpus callosum; a bundle of nerve fibers connecting the left and right hemispheres. It acts like a quality control manager for our thoughts and actions, constantly scanning for mismatches between what we intend to do and what we actually do. When the ACC detects a problem, such as realizing we've spoken harshly or broken a commitment, it sends a signal to other brain regions, especially the prefrontal cortex, to pause, evaluate, and make a course correction.

The ACC is deeply connected to emotional and social systems, meaning it doesn't just notice logical errors; it also flags when we've stepped outside our values or caused relational harm. In Step 10, this monitoring role is crucial, it helps us recognize when we've gone off track so we can take action before small problems grow into major setbacks.

Long-Term Resilience in the ACC and PFC

Practicing Step 10 daily doesn't just create short-term relief, it reshapes the brain's core regulation systems over the long haul. The anterior cingulate cortex (ACC) and prefrontal cortex (PFC), which work together to monitor, evaluate, and adjust behavior, both benefit from consistent use.

At first, this might require effort. The ACC must scan for discrepancies between our values and our actions, while the PFC steps in to plan a repair. But repetition builds efficiency and efficacy. Over time, these brain regions begin to respond more quickly and with less internal conflict, making accountability less of a struggle and more of an ingrained habit. Neurologically, this is similar to strengthening a muscle:

- *Initial phase:* The circuits are weak, and the process feels slow, resistant or uncomfortable
- *Adaptation phase:* The ACC detects mistakes earlier, and the PFC organizes corrective action with less overthinking
- *Maintenance phase:* Self-correction becomes automatic, requiring less conscious energy, which frees up mental resources for growth in other areas of life

Long-term resilience in these systems also means:

- *Reduced amygdala overreaction:* mistakes no longer trigger the same degree of fear, shame or avoidance
- *Faster emotional recovery:* the brain returns to baseline more quickly after stress or a triggering event
- *Increased mental flexibility:* openness to external feedback without defensive shutdown

With years of Step 10 practice, the ACC and PFC essentially "learn" that prompt accountability is safe, possible, and rewarding. This resilience is not just psychological, it's a structural and functional change in the brain's control systems that helps protect and reinforce recovery for life.

Emotional Regulation
Calming the Inner Alarm

Recognizing we've made a mistake can activate the amygdala, sparking anxiety, defensiveness, or shame. If left unchecked, this can lead to avoidance or dishonesty, reopening the door to old patterns. Step 10 helps rewire this process:

- *The ACC flags the issue*
- *The PFC steps in to evaluate and plan a corrective response*
- *The parasympathetic nervous system (rest-and-digest mode) is activated through conscious choice, slowing heart rate and calming the body*

Over time, this reduces the stress response linked to admitting mistakes and makes self-correction feel safer and more natural.

The Role of Neuroplasticity

Every time we notice an error and respond with accountability, we strengthen the neural pathways for honesty, impulse control, and emotional regulation. T his is neuroplasticity in action:

- *Old patterns of denial and defensiveness weaken*
- *New patterns of openness and prompt repair become the brain's default*
- *Just as physical exercise strengthens muscle, daily inventory strengthens the neural "muscles" of inner integrity.*

Mindfulness and the Default Mode Network

Step 10 benefits from mindfulness, the practice of nonjudgmental awareness of the present moment. Mindfulness reduces overactivity in the default mode network (DMN), the self-referential brain system that, when dysregulated, can spiral into rumination and self-justification. By regularly checking in with ourselves and observing our behavior without immediate defensiveness, we create space for course correction without self-condemnation.

Neurochemistry of Prompt Admission

When we promptly admit we're wrong and take corrective action, several neurochemical changes can occur:

- *Serotonin supports mood stability and cognitive flexibility during self-evaluation*
- *Dopamine can be released when we act in alignment with our values, reinforcing honesty as rewarding*
- *Oxytocin can increase when admissions lead to repaired or increased trust in relationships*

This chemical feedback loop helps make integrity feel good, encouraging repetition of the behavior.

Life Example
A Small Course Correction

Mei was three years sober when she noticed an old habit creeping back in. She had snapped at a coworker during a stressful afternoon and immediately felt her stomach twist with shame. Her old instinct was to pretend it hadn't happened and hope it would fade. But Step 10 had taught her a different rhythm. Instead of burying the mistake, she paused, breathed, and paid attention to the uncomfortable signal her body was giving her. "This is your reminder," she told herself, "Not your punishment." Integrity was becoming a habit rather than a struggle.

Before she left work, she walked over to her coworker's desk. "I was short with you earlier," she said. "You didn't deserve that. I'm sorry. "The coworker shrugged and smiled. "Rough day. It's okay."

The apology took ten seconds, but it shifted something deeper. Mei felt her chest loosen and her thoughts become clearer. She wasn't dragging a mistake behind her into the evening. She had corrected it—and in doing so, her brain reinforced the growing pathway that linked honesty with peace.

Why This Step Matters Neurologically

Step 10 is like daily maintenance for the brain's moral compass. It:

- *Keeps self-monitoring systems sharp*
- *Reduces defensive responses in the amygdala*
- *Strengthens prefrontal control over impulse and reactivity*

- *Creates neurochemical rewards for accountability*
- *Builds resilience against relapse by preventing negative emotional buildup*

By addressing small issues promptly, we avoid the backlog of guilt, resentment, and shame that can erode recovery.

Reflection Questions

How do you currently notice when you've acted outside your values?

What helps you respond promptly rather than avoid or justify?

How might a daily inventory help you keep relationships clear and healthy?

Could you create a "prompt repair" rule for yourself? What would it look like?

Which part/s of your brain gets "stronger" when you promptly admit mistakes?

Key Neurobiological Claims for Step 10

The anterior cingulate cortex (ACC) monitors for errors and value-based discrepancies.

Step 10 strengthens the ACC–PFC connection for prompt, thoughtful correction.

Prompt admission reduces stress reactivity.

Engaging the parasympathetic nervous system calms amygdala-driven defensiveness.

Mindfulness reduces overactive DMN patterns.

Minimizing rumination or justification helps create mental space for honest reflection.

Neuroplasticity strengthens integrity circuits.

Through repeated practices of self-correction, the circuits change to become more robust.

Neurochemicals like serotonin, dopamine, and oxytocin are key.

They help reinforce the positive emotional payoff of accountability.

In the next chapter, we'll look at Step 11: Spiritual Practice and the Brain's Attention Networks, exploring how prayer and meditation reshape neural pathways for focus, emotional balance, and a deeper sense of relatedness to the world around you.

Chapter 12
Step 11 – Spiritual Practice and the Brain's Attention Networks
How Prayer and Meditation Reshape Focus, Emotional Balance, and Cognitive Integration

Sought through prayer and meditation to improve our conscious contact with God as we understood Him, praying only for knowledge of His will for us and the power to carry that out.

Step Eleven of Alcoholics Anonymous

Step 11 invites us to deepen our recovery through spiritual practice, whether that means formal prayer, meditation, mindfulness, or another form of intentional stillness and intentional listening. Spiritually, it's about connection: with a higher power, with others, and with our truest selves. In essence, we connect with something bigger and better in order to reconnect with the best parts of our core human nature. Psychologically, there is an awakening to a sense of inner agency, an awareness that we can focus our thoughts and attention in a way that improves the quality of our life.

Neurobiologically, these practices strengthen attention networks, regulate emotional systems, and enhance brain connectivity. They can quiet the mental chatter that fuels anxiety or craving while supporting a grounded, open state of mind. Over time, these changes make it easier to live in alignment with our values and our closest communities, both of which promote a healthy life and help maintain recovery.

Attention Networks and Recovery

Two major brain systems help us focus and stay grounded during spiritual practice:

- *The Executive Attention Network: Located in the prefrontal cortex and parietal lobes, this network manages sustained attention and goal-directed thinking*
- *The Salience Network: Anchored in the anterior cingulate cortex (ACC) and insula, it determines what deserves our attention, shifting between internal and external focus*

In addiction, these systems can become biased toward salience of drug-related cues, wherein drugs and alcohol command our attention. Spiritual practices retrain these networks, helping the brain prioritize values-based goals over impulsive cravings.

Quieting the Default Mode Network (DMN)

The default mode network (DMN) is the brain's "background chatter" system, active during mind-wandering, daydreaming, or self-referential thought. In earlier chapters, we've seen the DMN's dual nature:

- *Positive role: Supports reflection, self-understanding, and meaning-making*
- *Negative role: Can over-activate in rumination, self-criticism, and worry*

Prayer and meditation balance DMN activity, keeping the reflective benefits while reducing unproductive rumination. Studies show that long-term meditators have more flexible DMN patterns, with quicker transitions to focused attention when needed.

Emotional Regulation and the Limbic System

Spiritual practice also affects the limbic system, the emotional core of the brain, including the amygdala, hippocampus, and hypothalamus. Regular meditation has been shown to:

- *Reduce amygdala reactivity to stress*
- *Increase hippocampal volume, supporting memory and emotional context*
- *Improve hypothalamic regulation of the stress response; hypothalamic–pituitary–adrenal (HPA) axis*

These changes translate into feeling calmer, less reactive, and better able to handle the ups and downs of life without using substances to balance our emotions or reactions.

Meditation, Hippocampal Volume, and Recovery

The hippocampus is a seahorse-shaped structure deep in the brain that plays a central role in memory formation, emotional context, and stress regulation. In people with substance use disorders, chronic stress and high cortisol levels can shrink hippocampal volume, making it harder to:

- *Accurately recall events without emotional distortion*
- *Contextualize current challenges within a broader perspective*
- *Regulate the stress response through the HPA axis*

This shrinkage can contribute to reactivity, impulsivity, and difficulty learning from past mistakes, all of which undermine recovery.

Why This Matters in Recovery

Structural MRI studies have shown that regular meditation can increase hippocampal volume over time. A larger, healthier hippocampus supports:

- *Better emotional context: the ability to see a current challenge without being overwhelmed by past trauma*
- *Improved memory accuracy: reducing the distortions caused by stress or craving states*
- *More effective stress regulation: helping prevent the cortisol surges that can lead to relapse triggers*

By supporting hippocampal health, meditation strengthens the brain's ability to respond to life's challenges with perspective and clarity instead of "blind" or automatic reactivity.

Meditation and Theta Waves

Brain activity is often described in terms of wave patterns, measured in hertz (Hz). Theta waves are slower brain rhythms, typically in the range of 4–8 Hz. They are most prominent during:

- *Light sleep and dream states*
- *Deep relaxation*
- *Certain meditative states*

Why Theta Waves Matter in Meditation and Recovery

Theta waves are associated with heightened creativity, intuition, and emotional integration. They facilitate memory consolidation, helping to reframe and integrate past experiences. In recovery, theta states can create a "bridge" between conscious awareness and subconscious emotional material, allowing for gentle processing of unresolved feelings.

Meditation can increase the amount of time the brain spends in this theta-dominant state, which may explain why people often emerge from meditation feeling refreshed, open, and more emotionally centered.

Bringing It Together

When we meditate regularly, we are not only calming the mind in the moment, we are literally reshaping the brain's structure and rhythms. The hippocampus grows healthier, supporting memory, context, and emotional regulation. Theta wave activity increases, promoting deeper self-integration and reduced reactivity. In the context of Step 11, these changes make "conscious contact" with a higher power not only a spiritual experience but also a neurobiological strengthening of the systems that protect long-term recovery.

Neurochemistry of Spiritual Practice

Several neurotransmitters are involved in the positive effects of prayer and meditation:

- *Serotonin: Enhances mood, promotes a sense of well-being, and supports flexible thinking*
- *Dopamine: Reinforces motivation to return to the practice by linking it to pleasure and reward*
- *GABA: Calms neural activity, reducing anxiety and physical tension*
- *Endorphins: Promote feelings of peace, connection, and even mild euphoria after deep practice*

This neurochemical profile supports both mental clarity and emotional steadiness, key to long-term recovery.

Life Example
Rebuilding Inner Stillness

Luis had grown accustomed to chaos. In active addiction, silence made him restless. Even in early recovery, sitting still felt like something dangerous might rise up inside him. Prayer and meditation sounded impossible. At his sponsor's suggestion, he experimented with two minutes a day—just sitting with his eyes closed and noticing his breath. His mind darted in every direction. Some mornings he felt frustrated. Other days he felt numb. But he kept trying.

After a few weeks, something shifted. He noticed he wasn't reacting as quickly when his kids spilled juice or when traffic slowed. He didn't feel that old surge of irritation shoot through his chest. When problems came up at work, he found himself thinking before speaking. He couldn't point to an exact moment of change; it was more like an inner landscape slowly softening. "It's like my mind has more room in it," he said. "I don't get trapped in every thought anymore." Step 11 didn't make life quiet—it made him quiet enough inside to navigate life with steadiness and clarity.

Why Step 11 Matters Neurologically

Step 11 integrates cognitive control, emotional regulation, and spiritual meaning-making, which:

- *Strengthens executive attention and salience networks*
- *Balances DMN activity for healthy reflection*
- *Calms limbic reactivity, reducing stress and craving triggers*
- *Reshapes hippocampal volume for healthier memory and emotional regulations*
- *Promotes brain theta waves for introspective processing and reduced emotional reactivity*
- *Builds neurochemical states that support resilience and connection*

Practiced daily, these changes accumulate, making it easier to stay present, resist impulsive behavior, and live in alignment with recovery principles.

Reflection Questions

How do you currently practice Step 11? Prayer, meditation, mindfulness, or another form of spiritual focus?

When you practice, what changes do you notice in your mood, attention, or reactivity?

What might keep you from committing to a small, consistent daily practice of prayer or meditation?

What does "conscious contact" mean for you in your own recovery, and do you have that "contact" outside of prayer or meditation?

What are the benefits of increasing hippocampal volume through meditation?

Key Neurobiological Claims for Step 11

Attention networks can be retrained.

Spiritual practices can shift salience away from addictive cues and toward values-based focus.

The DMN is balanced.

Shifting the orientation from rumination (aka: replaying old tapes) while preserving healthy, honest and accurate reflection.

The limbic system is calmed.

Reducing spikes or surges of amygdala-driven stress and emotional reactivity.

Neurochemical changes take place.

Serotonin, dopamine, GABA, and endorphins take effect to help support mood stability and feelings of wellness.

Daily repetition builds structural and functional brain changes.

These are the biological changes that strengthen recovery resilience.

Regular meditation can increase hippocampal volume.

This supports better emotional context, more accurate memory, and improved stress regulation, all of which improve mental resilience and can help reduce relapse risk in recovery.

Meditation enhances theta brain wave activity.

Fostering emotional integration, creativity, and subconscious processing, this can help reframe past experiences and promote calm, centered decision-making.

In the next chapter, we'll turn to Step 12: Service and the Brain's Altruism Circuits, exploring how helping others activates reward, empathy, and connection systems, making service not just a moral choice, but a neurologically sustaining part of recovery.

Chapter 13
Step 12 – Service and the Brain's Altruism Circuits
How Helping Others Strengthens Reward, Empathy, and Connection Systems

Having had a spiritual awakening as the result of these steps, we tried to carry this message to alcoholics, and to practice these principles in all our affairs.
Step Twelve of Alcoholics Anonymous

Step 12 is the culmination of the recovery process, where personal healing flows outward into service, mentorship, and living the principles of recovery. Spiritually, it's about gratitude and giving back. Psychologically, it reinforces identity as a person in recovery, capable of contributing value to others' lives. Neurobiologically, service activates the brain's altruism circuits, strengthening reward systems, empathy networks, and social bonding pathways. This not only benefits others, but it also cements the brain changes that recovery has been building all along.

The Brain's Reward in Giving

Acts of service engage the mesolimbic dopamine system, a network of brain regions that generates motivation and pleasure, and a network often hijacked by addictive substances. This pathway includes key structures like the ventral tegmental area (VTA) and ventral striatum.

Here's how they work together in the context of giving and being of service:

- *When we help someone in a way that feels meaningful, the VTA releases dopamine into the ventral striatum, producing feelings of satisfaction and motivating us to repeat the behavior*

- *The ventral striatum acts as a central hub, integrating dopamine signals with emotional and social context, helping the brain link this act of helping with a sense of reward*
- *Endorphins, the brain's natural opioids, are often released alongside dopamine during acts of giving, creating a sense of warmth, mild euphoria, and stress reduction*

This combination produces a multi-layered reward experience:

- *Dopamine drives the motivation to act*
- *The ventral striatum evaluates and reinforces the behavior as meaningful*
- *Endorphins smooth the edges, producing a "feel-good" state that deepens our emotional connection to the act of giving and service*

In recovery, this process is especially important because it reclaims the brain's reward circuits from addictive substances, associating pleasure and motivation with prosocial, life-affirming behaviors. Said differently, the brain begins to associate pleasure with purpose, reducing the pull of old addictive cues.

Oxytocin and Social Bonding

Helping others, especially in face-to-face contexts, can trigger the release of oxytocin, the "bonding hormone." Oxytocin:

- *Increases trust and empathy*
- *Reduces fear and defensiveness*
- *Strengthens group cohesion*

For someone in recovery, this means service not only supports others' healing, but it also deepens their own sense of belonging, a critical factor in long-term sobriety.

Empathy Networks in Action

Service work activates the social cognition network, particularly:

- *Temporoparietal Junction (TPJ): perspective-taking and understanding others' needs*
- *Medial Prefrontal Cortex (mPFC): moral reasoning, empathy, and integrating values into behavior*
- *Anterior Insula: sensing and resonating with others' emotional states*

Over time, repeated engagement in service strengthens these circuits, making compassionate responses more automatic. It begins to feel like a reinforcing, pro-social reflex.

The Helper's High

Neuroimaging studies show that giving can produce a "helper's high," a blend of dopamine, oxytocin, and endorphin release. This chemical cocktail not only feels good, it also:

- *Reinforces altruistic behavior through reward pathways*
- *Buffers against depression and isolation*
- *Strengthens resilience to stress*

In recovery, this high becomes a healthy replacement for the artificially induced highs of substance use. This makes biological sense as endorphins are one of the body's natural opioids.

Life Example
Finding Pleasure in Giving Back

Derrick never considered himself the "helpful type." During his addiction, every action was about survival, just getting through the day, managing withdrawal, protecting the next drink. Caring about others felt like a luxury he couldn't afford. But when he reached Step 12, his sponsor encouraged him to take on a simple service commitment: making coffee before his regular meeting. Derrick reluctantly agreed.

At first, he did it out of obligation. But over time, he started to notice small things he had never paid attention to: the way new members relaxed when someone greeted them, the quiet conversations that happened near the coffee pot, the gratitude people expressed for something as simple as a hot cup. One night, a newcomer approached him and said, "I came back because you remembered my name."

Driving home afterward, Derrick felt a warm, unexpected sense of purpose wash over him. It wasn't dramatic, but it was real. For the first time in years, he wanted to show up—not for alcohol, not for escape, but for people. And slowly, the thought of sponsorship seemed like a goal he could achieve. A goal he wanted to achieve.

Why Step 12 Matters Neurologically

Step 12 transforms recovery into a self-sustaining feedback loop:

- *Giving to others activates reward systems that make service intrinsically motivating*
- *Social bonding hormones deepen connection and belonging*
- *Empathy networks become more responsive and efficient with repeated use*

These neurological benefits are protective against relapse, as they anchor pleasure, purpose, and connection in healthy, life-affirming, community building activities.

Reflection Questions

How has helping others affected your own recovery?

What kinds of services feel most meaningful to you?

How might you integrate acts of service into daily life, not just within the recovery community?

Who helped you in your early recovery, and how could you pass that on?

What neurochemicals cause the "Helper's High" in the brain? What are their functions?

Key Neurobiological Claims for Step 12

Acts of service activate the brain's mesolimbic dopamine system.

The ventral tegmental area, nucleus accumbens, and ventral striatum are engaged and activated in a way that rewards and reinforces healthy, prosocial behavior.

Oxytocin release increases trust, empathy, and social cohesion.

Helping others in safe, intimate settings can trigger release of the so-called "bonding hormone," which in turn helps to strengthen social recovery networks.

Empathy-related brain region (TPJ, mPFC, anterior insula) activity.

This area of the brain grows more responsive through repeated use, making a commitment to compassionate and empathetic behavior feel more natural and reflexive.

The "helper's high" blends dopamine, oxytocin, and endorphins.

This combination creates a sustainable positive feedback loop that supports long-term sobriety. Endorphins are in fact endogenous opioids.

In our next section, we'll step back and integrate the neurobiological story behind all 12 steps, seeing how each contributes to a whole-brain recovery process, one that supports not only sobriety, but a fuller, richer life.

Part III
Integration and Practice

Chapter 14
Living Neurobiological Recovery
Integrating Brain, Spirit, and Community in Daily Life

Recovery is not just about stopping the use of alcohol or drugs, it's about living in a way that reshapes and reorients the brain, renews the body, and restores the spirit. Over the course of the 12 steps, we've seen how each practice interacts with different brain systems: the prefrontal cortex strengthening self-control, the limbic system learning new emotional responses, the default mode network balancing reflection with presence, and the social cognition networks deepening empathy and connection.

Now, in this final section, we bring these threads together. Neurobiological recovery is not something separate from spiritual, social, or psychological recovery. Rather, it is the biological foundation that allows those dimensions to flourish. The brain is the soil; the steps are the cultivation. What grows from that soil is a life marked by clarity.

The Brain as a System, Not Just Parts

Throughout this book, we've highlighted specific regions, like the anterior cingulate cortex, the hippocampus, or the temporoparietal junction, and their roles in different steps. But it's important to remember that the brain never works in isolation. It is a living network, where circuits and chemicals interact in real time, influenced by environment, relationships, and practice.

This means that recovery is not about "fixing" one broken area. Instead, it is about shaping patterns across the whole system:

- *Calming fear responses while strengthening pathways for courage*
- *Reducing self-centered rumination while enhancing networks for empathy*
- *Replacing dopamine-driven craving with dopamine-driven purpose*

When these changes take hold together, the brain becomes more integrated, and recovery begins to feel less like a battle and more like a natural rhythm of living.

Daily Practices that Reshape the Brain

The 12 steps are not one-time tasks; they are daily practices that reinforce healthy circuits again and again. Just as a musician strengthens skill through rehearsal, the recovering brain strengthens recovery through repetition. Each step plays its part, but together they create a self-reinforcing loop:

- *Honesty and humility reduce defensiveness, calming the amygdala*
- *Prayer and meditation increase hippocampal health and regulate the stress response*
- *Service work activates reward pathways, making altruism intrinsically satisfying*

Over time, these practices change not only the way we act, but the way we are. They rewire our baseline responses to stress, joy, fear, and connection.

Recovery as a Social Brain Experience

Humans are wired for relationships, and the brain is deeply social by design. Recovery communities take advantage of this fact. The presence of others who listen, guide, and share experience helps regulate our nervous systems. Oxytocin, mirror neurons, and social cognition networks all work together to create trust and empathy.

This is why isolation is so dangerous for someone with an addicted brain, and why fellowship, sponsorship, and service are so healing. The group is not just "support." It is a neurobiological regulator, helping each person hold balance they could not achieve alone. Step 12 makes this explicit, but the truth runs through the entire program: recovery is contagious, not just psychological but neurobiological.

Living with an Integrated Brain

When the prefrontal cortex, limbic system, and social networks begin to operate in harmony, recovery stops feeling like white-knuckled resistance and becomes more like integration. Integrity, in its deepest sense, means "wholeness." A recovering brain is a more integrated brain, one where self-control, emotional regulation, empathy, and spiritual reflection all work together.

The fruit of this integration is not simply sobriety. It is a renewed capacity for joy, maturity, and meaning. The spiritual awakening described in Step 12 is mirrored in the brain: new circuits, new chemistry, and a new way of relating to the world.

Moving Forward

Neurobiology cannot replace the spiritual, social, and psychological dimensions of recovery. But it can give us language for what is happening beneath the surface, helping people understand that change is not only possible, it is already happening in their cells and circuits.

Living neurobiological recovery means honoring both the mystery of spiritual awakening and the tangible science of brain change. It is embracing the fact that when we practice the steps, we are training our brains for a healthy, honest, responsible way of being.

Chapter 15

Practical Tools for Practitioners and Individuals

Applying Neurobiology to Support Recovery

The science of recovery is not meant to stay in textbooks. It belongs in meeting halls, counseling offices, treatment centers, and living rooms; anywhere people are walking the path of sobriety. Understanding how the brain works in recovery allows both practitioners and individuals to translate neurobiology into practical tools.

These tools don't replace the 12 steps, they illuminate them. They give us new ways of encouraging honesty, humility, connection, and service by showing how these practices reshape brain circuits. When people see their growth not just as "willpower" or "grace," but also as biological healing, hope deepens and motivation strengthens. It's another way to understand recovery as real progress.

For Practitioners

Supporting the Recovering Brain

Professionals in counseling, medicine, ministry, or peer recovery support can use brain-based insights to strengthen treatment and guidance.

Normalize Relapse as a Brain Process

Relapse is often framed as failure, but neurobiology reminds us it is part of the rewiring process. The prefrontal cortex is still regaining strength, and old limbic pathways are still tempting. Practitioners can help clients reframe relapse as a sign of brain healing in progress. It's early practice, not moral collapse.

Highlight the Role of Stress Regulation

Many relapses are triggered by stress. Explaining the amygdala's overactivation and the calming effects of breathing, prayer, or co-regulation with others makes stress management a tangible neurological goal.

Encourage Service as Medicine

By describing how dopamine and oxytocin are released in acts of service, practitioners can reframe Step 12 as more than moral duty; it's a biological treatment that rewards altruism and sustains recovery. It feels good for a reason.

Use Language of Integration

Practitioners can explain that recovery is not about eliminating one bad part of the brain, but about strengthening connections across the system. This destigmatizes the idea of a "broken brain" and reinforces hope for wholeness.

For Individuals

Daily Practices that Train the Brain

Recovery is strengthened when individuals understand that their daily practices are literally brain-training exercises.

Journaling and Reflection (ACC Activation)

Writing about daily behavior strengthens the anterior cingulate cortex, the brain's self-monitoring system, making self-correction easier over time.

Mindful Breathing and Meditation (PFC and Hippocampus Growth)

Even five minutes of daily practice can enlarge hippocampal volume and increase prefrontal control over stress reactivity and responses.

Sharing with Others (Oxytocin and Mirror Neurons)

Honest disclosure in meetings activates bonding chemistry and empathy circuits, reducing shame and isolation while strengthening connection.

Acts of Service (Dopamine and Ventral Striatum)

Helping another person in recovery shifts reward systems away from craving and self-centered rumination and toward purpose and community.

Sleep, Nutrition, and Exercise (Whole-Brain Resilience)

Lifestyle choices directly affect neurotransmitter balance: GABA, serotonin, and dopamine, making sobriety and emotional balance more stable.

When individuals recognize that these practices reshape circuits day by day, they begin to see recovery less as an abstract "program" and more as neural training.

The Language of Hope

Perhaps the greatest tool of all is language itself. When practitioners and peers can explain recovery as real changes in the brain, hope becomes concrete. Instead of "you'll just feel better if you keep going," the message becomes:

- *"Each time you pause to breathe, you are strengthening your prefrontal cortex"*
- *"When you serve, your brain's reward system learns to associate joy with connection"*
- *"By sharing honestly, your oxytocin system is calming your fear circuits"*

These statements empower individuals to see recovery not as something mystical or unreachable, but as something happening inside them in real time.

Integration of Science and Spirit

Practical tools work best when they honor both science and spirit. For many in recovery, prayer and meditation are spiritual lifelines. Explaining how these practices affect the brain does not diminish their sacredness, it enhances it. It shows that spirituality and biology are not at odds, but partners in healing. When practitioners and individuals together embrace this integrated view, recovery is supported at every level: psychological, social, spiritual, and neurobiological. No part of the human being is isolated.

Chapter 16
The Whole-Brain Story of the 12 Steps
From Fragmentation to Integration

Addiction is often described as a disease of the mind, body, and spirit. Neurobiology helps us see that it is also a disease of the brain's networks, where circuits meant for survival, reward, and connection are overtaken by substances and distorted by craving. What begins as relief-seeking becomes a cycle of compulsion. The prefrontal cortex loses control. The limbic system rules with fear and desire. The default mode network turns self-reflection into rumination and self-criticism. The social brain retreats into isolation.

The 12 steps interrupt this cycle not only spiritually and socially, but also neurologically. Each step strengthens a different aspect of the brain, working together toward integration. The steps are not simply moral practices or therapeutic exercises.

A Journey of Neural Transformation

In the early steps, recovery begins by acknowledging powerlessness. This is not weakness, but the prefrontal cortex relearning humility, surrendering the illusion of total control so that new circuits can form. The amygdala, once hypervigilant, begins to calm as the limbic system no longer bears the burden of survival alone.

Steps 2 and 3 bring belief and surrender, shifting the default mode network from isolated self-focus into connection with meaning, trust, and larger perspective. These steps transform rumination into reflection, self-criticism into openness, and fear into peace.

Steps 4 and 5 encourage honest self-examination and confession, activating the anterior cingulate cortex and social bonding chemistry. Painful memories stored in the hippocampus are not erased but reconsolidated, re-contextualized by new experiences of safety and acceptance. The nervous system learns that truth-telling no longer brings punishment, it brings release.

Steps 6 and 7 turn toward readiness and humility, loosening the grip of the ego. Here the default mode network finds balance: still a source of meaning and self-understanding, but no longer a prison of self-centeredness. The medial prefrontal cortex and temporoparietal junction, critical for empathy and perspective-taking, begin to anchor the self in relation to others, not just within itself. Serotonin pathways reinforce this openness and flexibility.

Steps 8 and 9 focus on repair. Guilt, once overwhelming and paralyzing, now activates brain regions for moral reasoning and social connection in constructive ways. The dopamine system, once enslaved to substances, is recruited into the service of reconciliation. Making amends releases GABA, serotonin, and even oxytocin; calming, uplifting, and rewarding the act of healing.

Steps 10 and 11 establish daily rhythms of self-monitoring and spiritual practice. The anterior cingulate cortex sharpens, catching missteps quickly and gently. The hippocampus grows, the prefrontal cortex strengthens, and meditation fosters theta waves that integrate conscious and unconscious material. The limbic system quiets, stress regulation improves, and recovery becomes more resilient.

Finally, Step 12 orients the whole brain outward. Service activates the mesolimbic dopamine system, ventral striatum, and oxytocin pathways, producing the "helper's high." Empathy networks expand, mirror neurons resonate, and recovery becomes not just individual healing but communal flourishing. The brain learns that giving, not grasping, is its most sustainable source of reward.

From Survival Brain to Meaning Brain

Addiction narrows the brain into survival mode, focused on the next drink, the next high, the next relief from pain and/or isolation. Recovery through the 12 steps expands the brain back into a meaning-making organ. Circuits once commandeered by substances are reclaimed for joy, purpose, love, and spiritual connection.

This is not a linear process. Relapse and setbacks are part of neural retraining. But every practice of honesty, surrender, reflection, and service lays down new pathways. Over time, the brain itself becomes less fragmented, more integrated, more whole.

The Spirit in the Synapse

Science alone cannot explain why recovery transforms lives so profoundly. Nor does it need to. For many, the 12 steps are first and foremost a successful spiritual path. Yet neurobiology shows us that spirituality and brain change are not separate forces, they are intertwined.

At the most basic level, every thought, feeling, or behavior relies on a synapse, the tiny gap between two brain cells (neurons) where chemical messengers, called neurotransmitters, carry signals. Addiction strengthens unhealthy pathways by flooding certain synapses with dopamine in ways that reinforce craving. Recovery, by contrast, weakens old addictive associations and builds strong alternative pathways for regulation and healthy reward. Each time someone practices honesty, meditation, amends, or service, new patterns of firing occur, and over time those synapses grow stronger. Old circuits associated with shame, fear, and compulsion weaken, while new circuits of trust, calm, and connection are reinforced.

This is why the steps don't just feel transformative; they are physically rewiring the brain at the synaptic level. The same prayer that deepens "conscious contact" also shapes new communication between neurons. The same act of service that fulfills a spiritual obligation also strengthens reward pathways that now associate joy with connection instead of substances. The language of faith and the language of neuroscience are describing the same miracle from different angles. Recovery is spiritual awakening and synaptic rewiring. It is grace and growth.

Living the Whole-Brain Story

To live in recovery is to live with a brain that is still healing, still reshaping, still learning new rhythms. The journey does not end with the steps, it continues in daily life, in relationships, in spiritual practice, in service. The whole-brain story of the 12 steps is this: what begins in brokenness can end in integration. What begins in isolation can end in connection. What begins as craving can end in joy. In every step, the addicted brain is not only learning to survive without substances, it is learning to thrive as a fully human brain, wired for love, meaning, and freedom.

Part IV
Appendices
Integration and Practices

Appendix A
Glossary of Key Neurobiological Terms

This glossary offers quick definitions of key terms used throughout the book, explained in the context of recovery.

Amygdala: An almond-shaped structure deep in the brain that detects threats and triggers fear or anxiety. In recovery, calming the amygdala helps prevent automatic overreactions to stress.

Anterior Cingulate Cortex (ACC): A brain region that monitors mistakes and helps us adjust behavior. Strengthened in Step 10's daily inventory, supporting honest self-correction.

Cortisol: A hormone released by the body during stress through the hypothalamic–pituitary–adrenal (HPA) axis. Cortisol helps mobilize energy and focus attention in the short term, but when chronically elevated reinforces fear, weakens self-regulation, and disrupts emotional balance.

Default Mode Network (DMN): A network of brain regions active during rest and self-reflection. In balance, it supports meaning-making and perspective; in excess, it fuels rumination and self-criticism.

Dopamine: A neurotransmitter that drives motivation and reward. Hijacked in addiction, but in recovery it is redirected toward purpose, connection, and service.

Endorphins: The brain's natural opioids, released during joy, exercise, and acts of service. They produce feelings of warmth and ease, helping buffer stress.

GABA (Gamma-Aminobutyric Acid): The brain's main calming neurotransmitter. It helps reduce anxiety and is boosted when guilt is resolved through confession and amends.

Hippocampus: A seahorse-shaped structure that stores memories and regulates stress. Shrinks during chronic stress or substance abuse, but can regrow through meditation, connection, and recovery practices.

Medial Prefrontal Cortex (mPFC): Involved in self-reflection, moral reasoning, and empathy. Strengthened by humility (Step 7) and service (Step 12). Helps balance self-focus with care for others.

Mirror Neurons: Specialized brain cells that fire both when we act and when we observe others acting. They underlie empathy and connection, especially in Steps 5 and 9.

Oxytocin: Sometimes called the "bonding hormone." It promotes trust, connection, and openness. Released when we share honestly, receive compassion, or serve others.

Parasympathetic Nervous System (PNS): The "rest and digest" system that calms the body after stress. Activated by surrender (Step 3), prayer, meditation, and safe social contact.

Prefrontal Cortex (PFC): The region of the brain responsible for planning, decision-making, impulse control, and evaluating long-term consequences. In addiction the PFC becomes weakened. In recovery, honesty, self-reflection, meditation, and service help strengthen PFC pathways, allowing for better emotional regulation, improved judgment, and greater alignment between values and behavior.

Serotonin: A neurotransmitter that promotes mood stability and openness. Supports humility, patience, and emotional balance in recovery.

Synapse: The tiny gap between two neurons where communication happens. Neurotransmitters carry signals across synapses. Addiction strengthens synapses tied to craving, while recovery practices weaken those and strengthen healthier connections, literally rewiring the brain.

Temporoparietal Junction (TPJ): A brain hub important for perspective-taking and empathy. Crucial for making amends (Step 9) and for service (Step 12).

Ventral Striatum: A reward hub that integrates dopamine with emotional meaning. In recovery, it shifts pleasure away from substances and toward connection, purpose, and service.

Appendix B
Summary Table of the 12 Steps and the Brain

Step	Spiritual/Practical Theme	Primary Brain Systems Involved
1. Powerlessness	Honesty, admitting lack of control	Prefrontal cortex (self-control), limbic system (fear regulation)
2. Belief	Hope, openness to help	Default Mode Network (Meaning making), prefrontal cortex
3. Surrender	Letting go, trust	Amygdala (fear), HPA axis (stress), parasympathetic nervous system
4. Inventory	Honest self-examinaton	Hippocampus (memory), ACC (monitoring)
5. Confession	Sharing truth with others	Oxytocin (bonding), mirror neurons (empathy)
6. Readiness	Willingness to change	ACC and prefrontal cortex (self-monitoring)
7. Humility	Ego dissolution	DMN (reflection vs. rumination), serotonin, TPJ, mPFC

8. Repair	Preparing amends	TPJ (perspective-taking) mPFC (moral reasoning
9. Amends	Direct reconciliation	GABA (calming guilt), serotonin, dopamine, oxytoxin
10. Daily Inventory	Ongoing self-correction	ACC (error detection), PFC (long-term resilience)
11. Prayer & Meditation	Conscious contact, gudiance	Hippocampus (growth), PFC (focus), theta waves
12. Service	Carrying the message	Mesolimbic dopamine system, ventral striatum, oxytocin, empathy networks

Appendix C
Neurobiological Exercises for Recovery

These exercises can be woven into daily life to strengthen the brain circuits most impacted by addiction.

Breathing for Stress Regulation
Step 3: Surrender

Slow, deep breathing activates the parasympathetic nervous system, calming the amygdala and lowering stress hormones. Try the "4-7-8" method: inhale 4 seconds, hold 7, exhale 8.

Journaling for Self-Monitoring
Step 10: Daily Inventory

Writing activates the ACC, supporting self-awareness. End each day by asking: What did I do well? Where did I fall short? What can I do differently tomorrow?

Memory Reconsolidation through Reframing
Step 4: Inventory

Revisit a painful memory with a trusted listener, then add a new truth: That was painful, but today I am safe and supported. This helps the hippocampus update old emotional memories.

Meditation for Growth
Step 11: Prayer and Meditation

Just 5 minutes of focused breathing and returning attention when it wanders can increase hippocampal volume, strengthen prefrontal control, and promote calming theta waves.

Empathy through Active Listening
Steps 5 & 8: Confession & Repair

In conversation, repeat back what you heard before responding: "I hear you felt hurt when I…" This strengthens mirror neurons, the TPJ, and the mPFC, deepening empathy.

Acts of Service for Reward Rewiring
Step 12: Service

Helping others activates dopamine and oxytocin, producing the "helper's high." Daily acts of service retrain reward pathways to associate joy with connection rather than substances.

Gratitude Practice for Mood Balance
Step 7: Humility

Listing 3 things you are grateful for each day helps boost serotonin, shifting the brain away from self-centered rumination and toward contentment.

Appendix D
Recovery Understood by Brain Region

Frontal Control and Self-Monitoring Regions

Prefrontal Cortex (PFC): Seat of decision-making, planning, and impulse control. Impaired in addiction, strengthened in recovery.

Medial Prefrontal Cortex (mPFC): Involved in self-reflection, empathy, and moral reasoning.

Anterior Cingulate Cortex (ACC): Monitors mistakes, supports self-correction, and underlies readiness to change.

Limbic System
Emotion and Memory

Amygdala: Detects threats, triggers fear and anxiety. Often overactive in addiction, calmed through surrender and trust.

Hippocampus: Stores memory and context and regulates stress. Damaged by chronic substance use but grows back with meditation and connection.

Reward and Motivation Circuits

Ventral Striatum (including Nucleus Accumbens): Central hub for motivation and reward learning; appropriated by substances, reoriented to healthy rewards in recovery.

Mesolimbic Dopamine System: Pathway linking VTA to nucleus accumbens and prefrontal cortex, driving motivation and reinforcing behavior.

Networks for Meaning and Perspective

Default Mode Network (DMN): Active during self-reflection. Balanced DMN supports meaning-making; overactive DMN fuels rumination and self-centeredness.

Temporoparietal Junction (TPJ): Key for perspective-taking, empathy, and social awareness.

Regulation and Connection Systems

Parasympathetic Nervous System (PNS): Calms the body, counterbalances stress response (HPA axis).

Hypothalamic-Pituitary-Adrenal (HPA) Axis: Stress hormone system, overactivated in addiction, soothed in recovery.

Mirror Neurons: Fire during both action and observation; form the basis of empathy and resonance in human connection.

Appendix E
*Locations of Key Brain Regions**

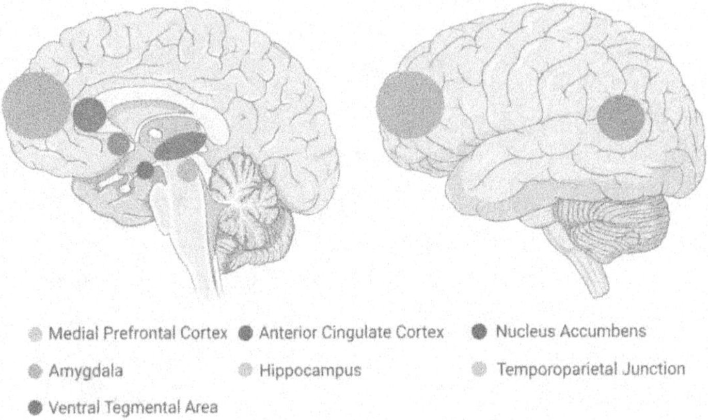

Approximate locations of selected cortical and subcortical regions discussed in the text, shown on medial (left) and lateral (right) views.

* *Labeled brain regions are intended for conceptual illustration rather than anatomical precision.*

Closing Note on the Appendices

Like all earlier sections of this book, these appendices are designed to complement the 12 steps, not replace them. They provide a biological lens for understanding why recovery practices work and offer practical ways to reinforce brain healing day by day.

www.ingramcontent.com/pod-product-compliance
Lightning Source LLC
LaVergne TN
LVHW012313020926
847253LV00043B/1177
9798897314669